OXFORD **REVISION** GUIDES

A Level

Advanced

HUMAN BIOLOGY

through diagrams

W R Pickering

Oxford University Press

Oxford University Press, Great Clarendon Street, Oxford OX2 6DP

Oxford New York
Athens Auckland Bangkok Bogota Bombay
Buenos Aires Calcutta Cape Town Dar es Salaam Delhi
Florence Hong Kong Istanbul Karachi
Kuala Lumpur Madras Madrid Melbourne
Mexico City Nairobi Paris Singapore
Taipei Tokyo Toronto Warsaw

and associated companies in
Berlin Ibadan

Oxford is a trade mark of Oxford University Press

First published 1998

ISBN 0 19 914192 4

Typesetting, design and illustration by Hardlines, Charlbury, Oxford
Printed in Great Britain

CONTENTS

Plant and animal cells

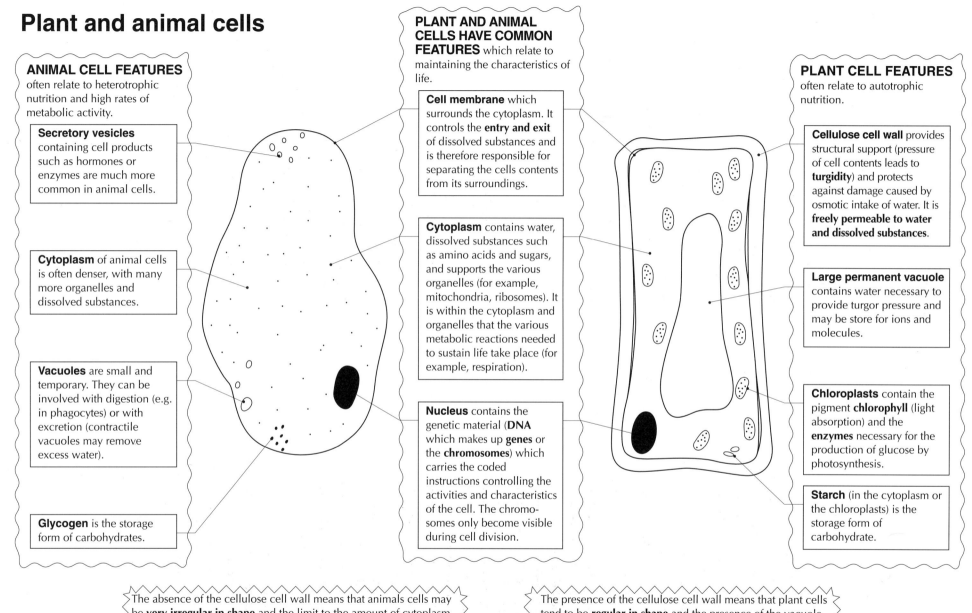

PLANT AND ANIMAL CELLS HAVE COMMON FEATURES which relate to maintaining the characteristics of life.

ANIMAL CELL FEATURES often relate to heterotrophic nutrition and high rates of metabolic activity.

Secretory vesicles containing cell products such as hormones or enzymes are much more common in animal cells.

Cytoplasm of animal cells is often denser, with many more organelles and dissolved substances.

Vacuoles are small and temporary. They can be involved with digestion (e.g. in phagocytes) or with excretion (contractile vacuoles may remove excess water).

Glycogen is the storage form of carbohydrates.

Cell membrane which surrounds the cytoplasm. It controls the **entry and exit** of dissolved substances and is therefore responsible for separating the cells contents from its surroundings.

Cytoplasm contains water, dissolved substances such as amino acids and sugars, and supports the various organelles (for example, mitochondria, ribosomes). It is within the cytoplasm and organelles that the various metabolic reactions needed to sustain life take place (for example, respiration).

Nucleus contains the genetic material (**DNA** which makes up **genes** or the **chromosomes**) which carries the coded instructions controlling the activities and characteristics of the cell. The chromosomes only become visible during cell division.

PLANT CELL FEATURES often relate to autotrophic nutrition.

Cellulose cell wall provides structural support (pressure of cell contents leads to **turgidity**) and protects against damage caused by osmotic intake of water. It is **freely permeable to water and dissolved substances**.

Large permanent vacuole contains water necessary to provide turgor pressure and may be store for ions and molecules.

Chloroplasts contain the pigment **chlorophyll** (light absorption) and the **enzymes** necessary for the production of glucose by photosynthesis.

Starch (in the cytoplasm or the chloroplasts) is the storage form of carbohydrate.

The absence of the cellulose cell wall means that animals cells may be **very irregular in shape** and the limit to the amount of cytoplasm which can be controlled by the nucleus means that animal cells may be **quite small** - about 25 µm diameter.

The presence of the cellulose cell wall means that plant cells tend to be **regular in shape** and the presence of the vacuole means that plant cells may be **quite large** - often 60 µm (or 0.06 mm) in diameter.

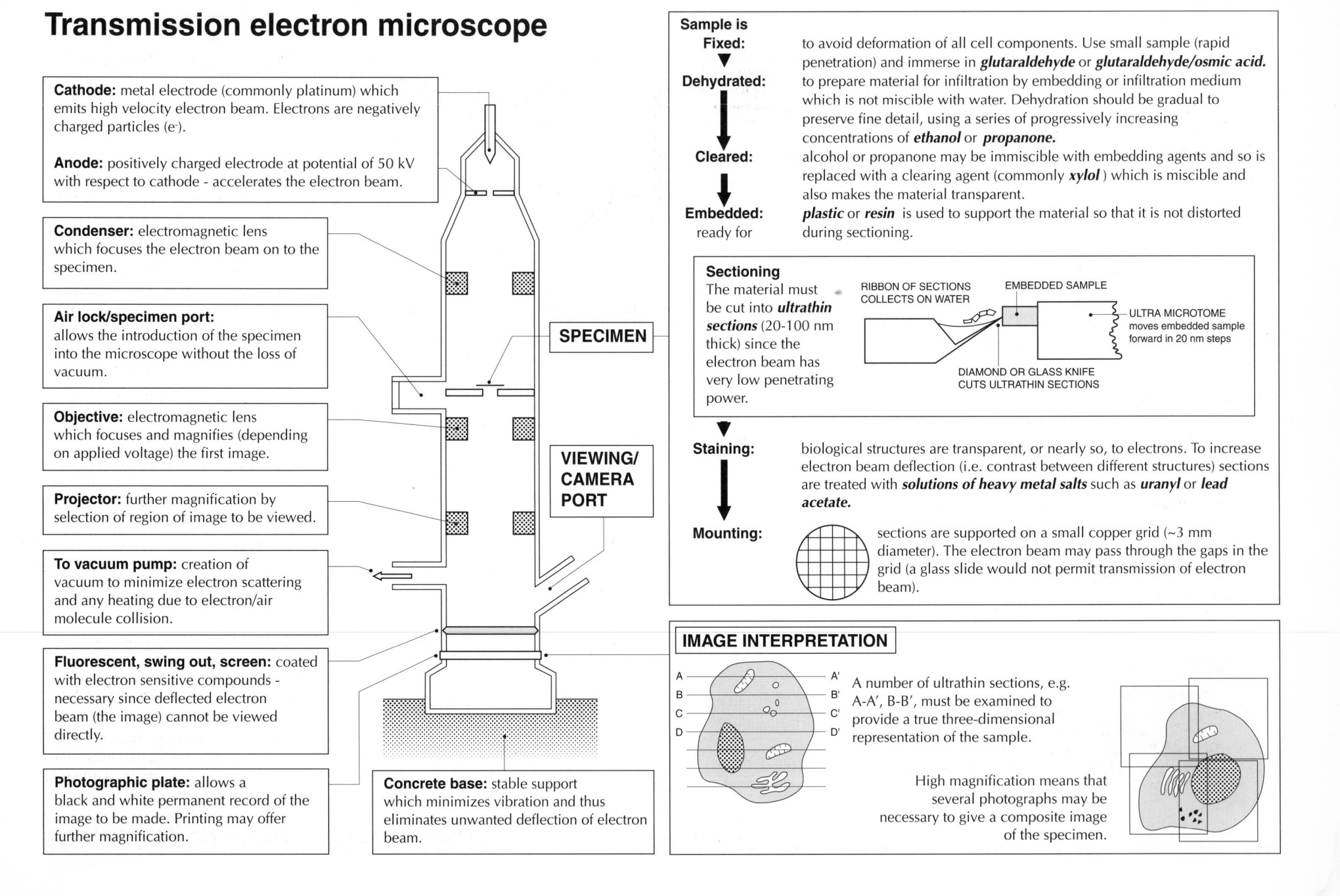

Transmission electron microscope

Cathode: metal electrode (commonly platinum) which emits high velocity electron beam. Electrons are negatively charged particles (e⁻).

Anode: positively charged electrode at potential of 50 kV with respect to cathode - accelerates the electron beam.

Condenser: electromagnetic lens which focuses the electron beam on to the specimen.

Air lock/specimen port: allows the introduction of the specimen into the microscope without the loss of vacuum.

Objective: electromagnetic lens which focuses and magnifies (depending on applied voltage) the first image.

Projector: further magnification by selection of region of image to be viewed.

To vacuum pump: creation of vacuum to minimize electron scattering and any heating due to electron/air molecule collision.

Fluorescent, swing out, screen: coated with electron sensitive compounds - necessary since deflected electron beam (the image) cannot be viewed directly.

Photographic plate: allows a black and white permanent record of the image to be made. Printing may offer further magnification.

Concrete base: stable support which minimizes vibration and thus eliminates unwanted deflection of electron beam.

SPECIMEN

VIEWING/ CAMERA PORT

Sample is Fixed: to avoid deformation of all cell components. Use small sample (rapid penetration) and immerse in **glutaraldehyde** or **glutaraldehyde/osmic acid.**

Dehydrated: to prepare material for infiltration by embedding or infiltration medium which is not miscible with water. Dehydration should be gradual to preserve fine detail, using a series of progressively increasing concentrations of **ethanol** or **propanone.**

Cleared: alcohol or propanone may be immiscible with embedding agents and so is replaced with a clearing agent (commonly **xylol**) which is miscible and also makes the material transparent.

Embedded: **plastic** or **resin** is used to support the material so that it is not distorted ready for during sectioning.

Sectioning
The material must be cut into **ultrathin sections** (20-100 nm thick) since the electron beam has very low penetrating power.

RIBBON OF SECTIONS COLLECTS ON WATER

EMBEDDED SAMPLE

ULTRA MICROTOME moves embedded sample forward in 20 nm steps

DIAMOND OR GLASS KNIFE CUTS ULTRATHIN SECTIONS

Staining: biological structures are transparent, or nearly so, to electrons. To increase electron beam deflection (i.e. contrast between different structures) sections are treated with **solutions of heavy metal salts** such as **uranyl** or **lead acetate.**

Mounting: sections are supported on a small copper grid (~3 mm diameter). The electron beam may pass through the gaps in the grid (a glass slide would not permit transmission of electron beam).

IMAGE INTERPRETATION

A number of ultrathin sections, e.g. A-A', B-B', must be examined to provide a true three-dimensional representation of the sample.

High magnification means that several photographs may be necessary to give a composite image of the specimen.

Scanning electron microscope

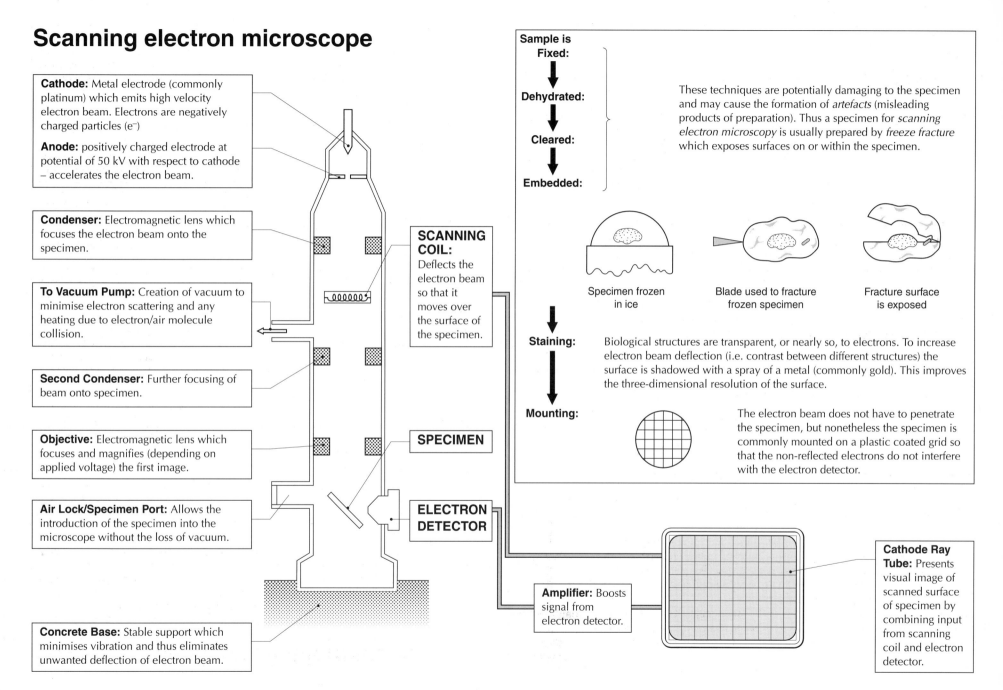

Cathode: Metal electrode (commonly platinum) which emits high velocity electron beam. Electrons are negatively charged particles (e^-)

Anode: positively charged electrode at potential of 50 kV with respect to cathode – accelerates the electron beam.

Condenser: Electromagnetic lens which focuses the electron beam onto the specimen.

To Vacuum Pump: Creation of vacuum to minimise electron scattering and any heating due to electron/air molecule collision.

Second Condenser: Further focusing of beam onto specimen.

Objective: Electromagnetic lens which focuses and magnifies (depending on applied voltage) the first image.

Air Lock/Specimen Port: Allows the introduction of the specimen into the microscope without the loss of vacuum.

Concrete Base: Stable support which minimises vibration and thus eliminates unwanted deflection of electron beam.

SCANNING COIL: Deflects the electron beam so that it moves over the surface of the specimen.

SPECIMEN

ELECTRON DETECTOR

Sample is
Fixed:

Dehydrated:

Cleared:

Embedded:

These techniques are potentially damaging to the specimen and may cause the formation of *artefacts* (misleading products of preparation). Thus a specimen for *scanning electron microscopy* is usually prepared by *freeze fracture* which exposes surfaces on or within the specimen.

Specimen frozen in ice

Blade used to fracture frozen specimen

Fracture surface is exposed

Staining:

Biological structures are transparent, or nearly so, to electrons. To increase electron beam deflection (i.e. contrast between different structures) the surface is shadowed with a spray of a metal (commonly gold). This improves the three-dimensional resolution of the surface.

Mounting:

The electron beam does not have to penetrate the specimen, but nonetheless the specimen is commonly mounted on a plastic coated grid so that the non-reflected electrons do not interfere with the electron detector.

Amplifier: Boosts signal from electron detector.

Cathode Ray Tube: Presents visual image of scanned surface of specimen by combining input from scanning coil and electron detector.

Animal cell ultrastructure

Microvilli are extensions of the plasmamembrane which increase the cell surface area. They are commonly abundant in cells with a high absorptive capacity, such as *hepatocytes* or cells of the *first coiled tubule of the nephron*. Collectively the microvilli represent a *brush border* to the cell.

Peroxisome is one of the group of vesicles known as *microbodies*. Each of them contains oxidative enzymes such as *catalase*, and they are particularly important in delaying cell ageing.

Lysosomes are sacs that contain high concentrations of hydrolytic (digestive) enzymes. These enzymes are kept apart from the cell contents which they would otherwise destroy, and they are kept inactive by an alkaline environment within the lysosome. They are especially abundant in cells with a high phagocytic activity, such as some *neutrophils*.

Centrioles are a pair of structures, held at right angles to one another, which act as organizers of the nuclear spindle in preparation for the separation of chromosomes or chromatids during nuclear division.

Free ribosomes are the sites of protein synthesis, principally for proteins destined for intracellular use. There may be 50 000 or more in a typical eukaryote cell.

Secretory vesicle undergoing exocytosis. May be carrying a synthetic product of the cell (such as a protein packaged at the Golgi body) or the products of degradation by lysosomes. Secretory vesicles are abundant in cells with a high synthetic activity, such as the cells of the *Islets of Langerhans*.

Endocytic vesicle may contain molecules or structures too large to cross the membrane by active transport or diffusion.

Smooth endoplasmic reticulum is a series of flattened sacs and sheets that are the sites of synthesis of steroids and lipids.

Microtubules are hollow tubes of the protein *tubulin*, about 25 nm in diameter. They are involved in intracellular transport (e.g. the movement of mitochondria), have a structural role as part of the cytoskeleton and are components of other specialized structures such as the centrioles and the basal bodies of cilia and flagella.

Rough endoplasmic reticulum is so-called because of the many ribosomes attached to its surface. This intracellular membrane system aids cell compartmentalization and transports proteins synthesized at the ribosomes towards the Golgi bodies for secretory packaging.

Nucleus is the centre of the regulation of cell activities since it contains the hereditary material, DNA, carrying the information for protein synthesis. The DNA is bound up with histone protein to form chromatin. The nucleus contains one or more nucleoli in which ribosome subunits, ribosomal RNA, and transfer RNA are manufactured. The nucleus is surrounded by a double nuclear membrane, crossed by a number of nuclear pores. The nucleus is continuous with the endoplasmic reticulum. There is usually only one nucleus per cell, although there may be many in very large cells such as those of striated (skeletal) muscle. Such multinucleate cells are called coenocytes.

Golgi apparatus consists of a stack of sacs called *cisternae*. It modifies a number of cell products delivered to it, often enclosing them in vesicles to be secreted. Such products include trypsinogen (from *pancreatic acinar cells*), insulin (from *beta-cells of the Islets of Langerhans*) and mucin (from *goblet cells in the trachea*). The Golgi is also involved in lipid modification in cells of the ileum, and plays a part in the formation of lysosomes.

Mitochondrion (pl. mitochondria) is the site of aerobic respiration. Mitochondria have a highly folded inner membrane which supports the proteins of the electron transport chain responsible for the synthesis of ATP by oxidative phosphorylation. The mitochondrial matrix contains the enzymes of the TCA cycle, an important metabolic 'hub'. These organelles are abundant in cells which are physically *(skeletal muscle)* and metabolically *(hepatocytes)* active.

Cytoplasm is principally water, with many solutes including glucose, proteins and ions. It is permeated by the *cytoskeleton*, which is the main architectural support of the cell.

Microfilaments are threads of the protein *actin*. They are usually situated in bundles just beneath the cell surface and play a role in endo- and exocytosis, and possibly in cell motility.

Plasmalemma (plasmamembrane) is the surface of the cell and represents its contact with its environment. It is differentially permeable and regulates the movement of solutes between the cell and its environment. There are many specializations of the membrane, often concerning its protein content.

A prokaryotic cell (e.g. a bacterium) has no true organelles.

* Important comparisons with eukaryotic cells.

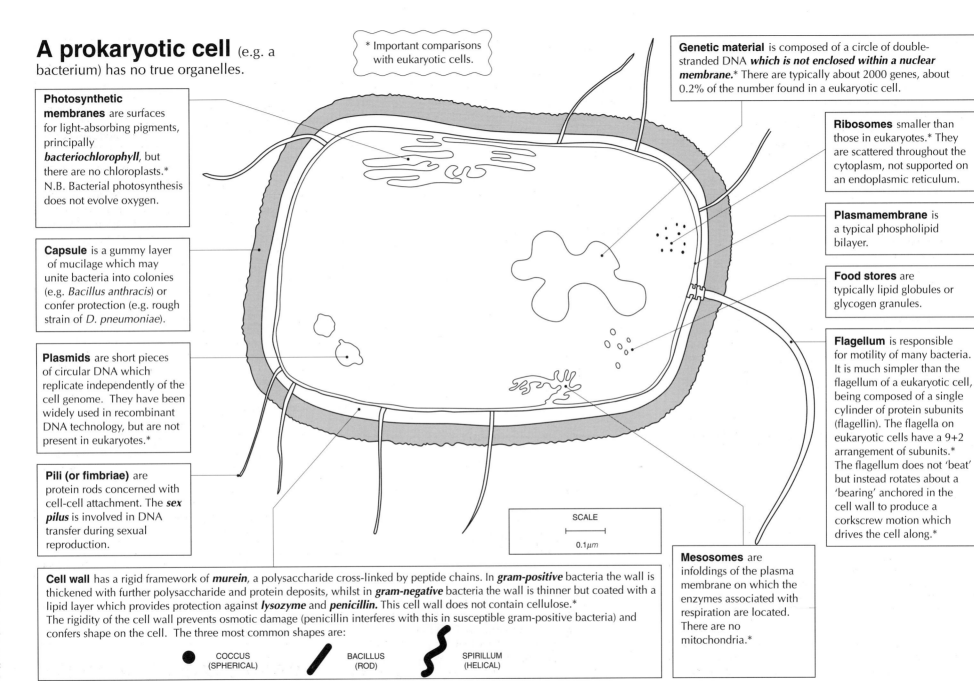

Photosynthetic membranes are surfaces for light-absorbing pigments, principally ***bacteriochlorophyll***, but there are no chloroplasts.* N.B. Bacterial photosynthesis does not evolve oxygen.

Capsule is a gummy layer of mucilage which may unite bacteria into colonies (e.g. *Bacillus anthracis*) or confer protection (e.g. rough strain of *D. pneumoniae*).

Plasmids are short pieces of circular DNA which replicate independently of the cell genome. They have been widely used in recombinant DNA technology, but are not present in eukaryotes.*

Pili (or fimbriae) are protein rods concerned with cell-cell attachment. The ***sex pilus*** is involved in DNA transfer during sexual reproduction.

Genetic material is composed of a circle of double-stranded DNA ***which is not enclosed within a nuclear membrane.**** There are typically about 2000 genes, about 0.2% of the number found in a eukaryotic cell.

Ribosomes smaller than those in eukaryotes.* They are scattered throughout the cytoplasm, not supported on an endoplasmic reticulum.

Plasmamembrane is a typical phospholipid bilayer.

Food stores are typically lipid globules or glycogen granules.

Flagellum is responsible for motility of many bacteria. It is much simpler than the flagellum of a eukaryotic cell, being composed of a single cylinder of protein subunits (flagellin). The flagella on eukaryotic cells have a 9+2 arrangement of subunits.* The flagellum does not 'beat' but instead rotates about a 'bearing' anchored in the cell wall to produce a corkscrew motion which drives the cell along.*

SCALE

0.1μm

Mesosomes are infoldings of the plasma membrane on which the enzymes associated with respiration are located. There are no mitochondria.*

Cell wall has a rigid framework of ***murein***, a polysaccharide cross-linked by peptide chains. In ***gram-positive*** bacteria the wall is thickened with further polysaccharide and protein deposits, whilst in ***gram-negative*** bacteria the wall is thinner but coated with a lipid layer which provides protection against ***lysozyme*** and ***penicillin.*** This cell wall does not contain cellulose.*
The rigidity of the cell wall prevents osmotic damage (penicillin interferes with this in susceptible gram-positive bacteria) and confers shape on the cell. The three most common shapes are:

COCCUS (SPHERICAL)   BACILLUS (ROD)   SPIRILLUM (HELICAL)

Differential centrifugation may be used to isolate cell components.

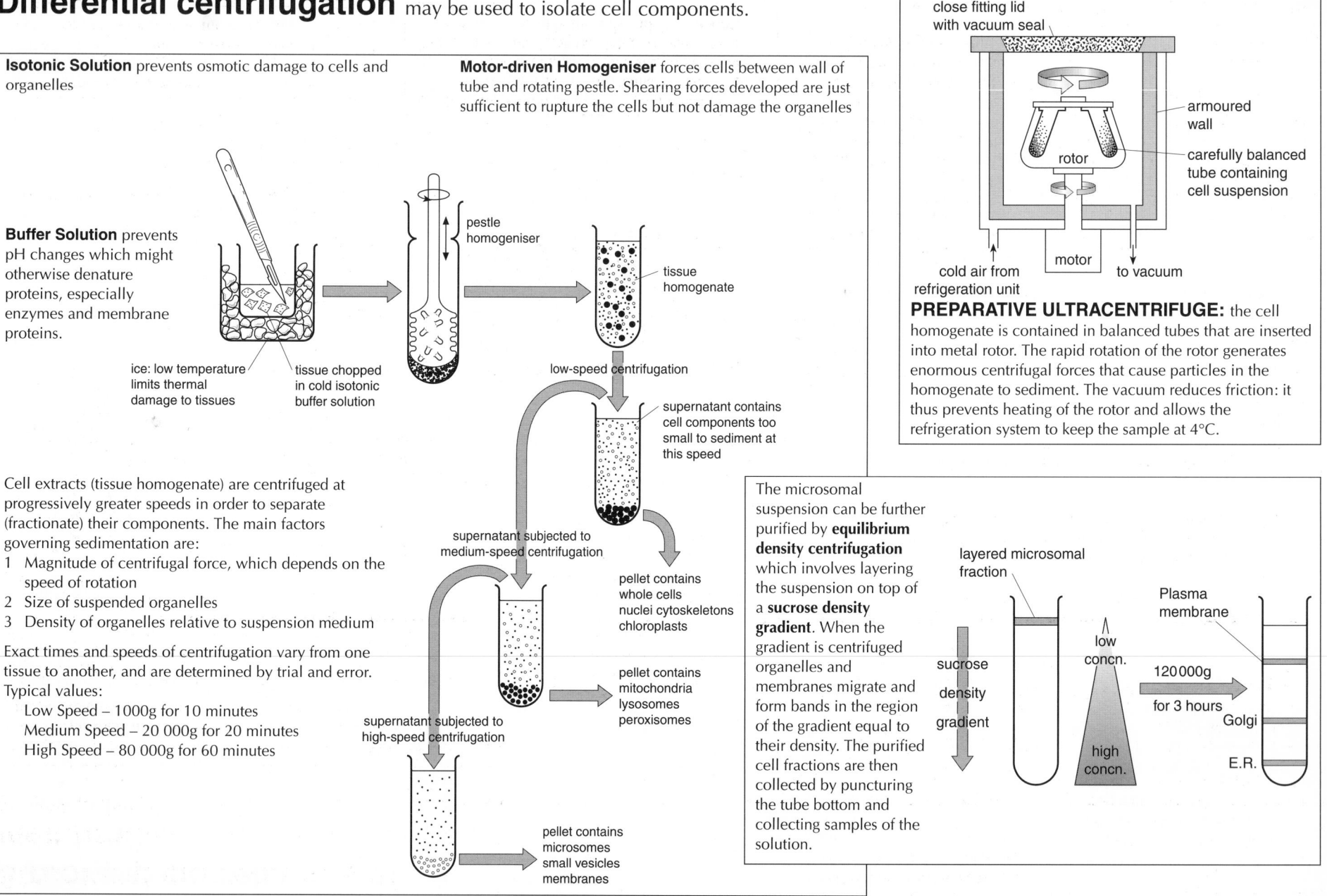

Isotonic Solution prevents osmotic damage to cells and organelles

Buffer Solution prevents pH changes which might otherwise denature proteins, especially enzymes and membrane proteins.

Motor-driven Homogeniser forces cells between wall of tube and rotating pestle. Shearing forces developed are just sufficient to rupture the cells but not damage the organelles

pestle homogeniser

tissue homogenate

ice: low temperature limits thermal damage to tissues

tissue chopped in cold isotonic buffer solution

low-speed centrifugation

supernatant contains cell components too small to sediment at this speed

Cell extracts (tissue homogenate) are centrifuged at progressively greater speeds in order to separate (fractionate) their components. The main factors governing sedimentation are:

1 Magnitude of centrifugal force, which depends on the speed of rotation
2 Size of suspended organelles
3 Density of organelles relative to suspension medium

Exact times and speeds of centrifugation vary from one tissue to another, and are determined by trial and error. Typical values:

  Low Speed – 1000g for 10 minutes
  Medium Speed – 20 000g for 20 minutes
  High Speed – 80 000g for 60 minutes

supernatant subjected to medium-speed centrifugation

pellet contains whole cells nuclei cytoskeletons chloroplasts

pellet contains mitochondria lysosomes peroxisomes

supernatant subjected to high-speed centrifugation

pellet contains microsomes small vesicles membranes

close fitting lid with vacuum seal

armoured wall

rotor

carefully balanced tube containing cell suspension

cold air from refrigeration unit

motor

to vacuum

PREPARATIVE ULTRACENTRIFUGE: the cell homogenate is contained in balanced tubes that are inserted into metal rotor. The rapid rotation of the rotor generates enormous centrifugal forces that cause particles in the homogenate to sediment. The vacuum reduces friction: it thus prevents heating of the rotor and allows the refrigeration system to keep the sample at 4°C.

The microsomal suspension can be further purified by **equilibrium density centrifugation** which involves layering the suspension on top of a **sucrose density gradient**. When the gradient is centrifuged organelles and membranes migrate and form bands in the region of the gradient equal to their density. The purified cell fractions are then collected by puncturing the tube bottom and collecting samples of the solution.

layered microsomal fraction

sucrose density gradient

low concn.

high concn.

120 000g for 3 hours

Plasma membrane

Golgi

E.R.

Structural components of membranes permit fluidity, selective transport and recognition, integrity and compartmentalization.

Because of the different solubility properties of the two ends of phospholipid molecules …

polar, so very soluble in water

non-polar, so very insoluble in water

… such molecules form a layer at a water surface

and a **phospholipid bilayer** can act as a barrier between two aqueous environments.

WATER

WATER

Hydrophilic heads point outwards: form hydrogen bonds with water

Hydrophobic tails point towards one another: this maximizes hydrophobic attractions and excludes water

Diffusion through aqueous channels in pore proteins: transmembrane proteins may have aqueous channels through which charged molecules may pass and thus avoid the hydrophobic tails of the phospholipid molecules.

Na$^+$

Some channels are open all of the time, but others are **gated** (they open and close only in response to a stimulus, such as a change in the membrane's electrical potential). Such **gated channels** are vital to the operation of nerve and muscle, where movements of Na$^+$, K$^+$ and Ca^{2+} initiate information transfer.

Diffusion across the lipid bilayer is responsible for the movement of **small, uncharged molecules.**

Thus O_2, H_2O, CO_2, urea and ethanol cross rapidly (they 'squeeze between') the polar phospholipid heads then dissolve in the lipid on one side of the membrane and emerge on the other.

Large or **charged molecules** cannot cross the lipid bilayer.

Thus Na$^+$, K$^+$, Cl$^-$, HCO$_3^-$ and glucose do not cross in this way.

Surface carbohydrates (collectively the **glycocalyx**) are usually oligosaccharides which are positioned to aid in cell recognition functions.

Active transport uses a **carrier protein** to transport a solute across a membrane but **energy is required** since transport may be **against a concentration gradient.** Typically ATP is hydrolysed and the binding of the phosphate group to the carrier changes the protein's conformation in such a way that the solute molecule is moved across the membrane.

Facilitated diffusion uses a **carrier protein** to transfer a molecule across a membrane **along** its electrochemical gradient. The binding of the solute alters the conformation of the carrier so that its position in the membrane changes and the solute molecule is discharged on the other side of the membrane. Glucose uptake by erythrocytes occurs in this way.

SOLUTE BINDING

CARRIER INVERSION

SOLUTE RELEASE AND CARRIER RETURN

N.B. There is **no requirement for ATP**, as there is **no energy consumption.**

Cell adhesion proteins firmly attach adjacent cells to one another, this is particularly important in epithelia. These proteins also serve as internal anchorage points for protein tubules of the cytoskeleton.

Lipid composition influences membrane fluidity: unsaturated fatty acid tails are 'kinked', limit close packing of the hydrophobic tails and so **increase** fluidity, but cholesterol may interfere with lateral movement of hydrophobic tails and thus **reduce** membrane fluidity.

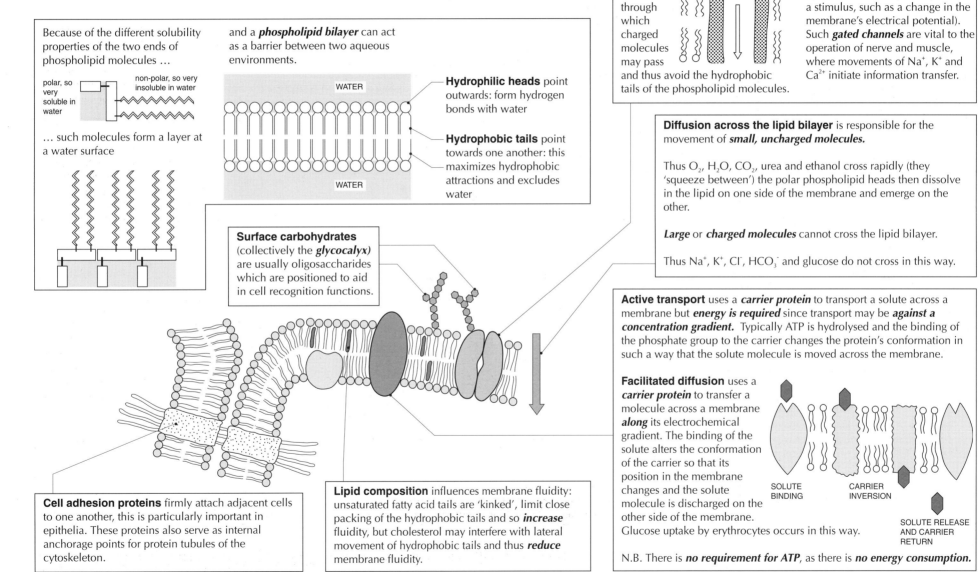

Diffusion, osmosis and active transport are processes by which

molecules are moved. Diffusion and osmosis are passive, but active transport requires energy.

DIFFUSION:

the movement of ions or molecules down a concentration gradient i.e. from a region of higher concentration to one of lower concentration.

This is a physical process which depends on the energy possessed by the molecules, thus
- small molecules diffuse faster than large molecules
- diffusion speeds up as temperature increases.

FICK'S LAW states that

Rate of diffusion is proportional to

$$\frac{SURFACE\ AREA \times DIFFERENCE\ IN\ CONCENTRATION}{THICKNESS\ OF\ MEMBRANE}$$

For living cells the principle (the movement of molecules down a concentration gradient) is the same, but there is one problem

→ the cell is

surrounded by a **cell membrane** which can restrict the free movement of the molecules

Important examples
- Oxygen from air sacs in the lung to blood, and from blood to cells

- Soluble foods from gut to blood

- Carbon dioxide from air to spaces inside leaf

→ This is **a selectively permeable membrane:** the composition of the membrane (lipid and protein) allows some molecules to cross with ease, but others with difficulty or not at all. In this example the membrane is permeable to water ● but not to the larger sucrose molecule ▨ - the simplest sort of selection is based on the **size** of the molecules.

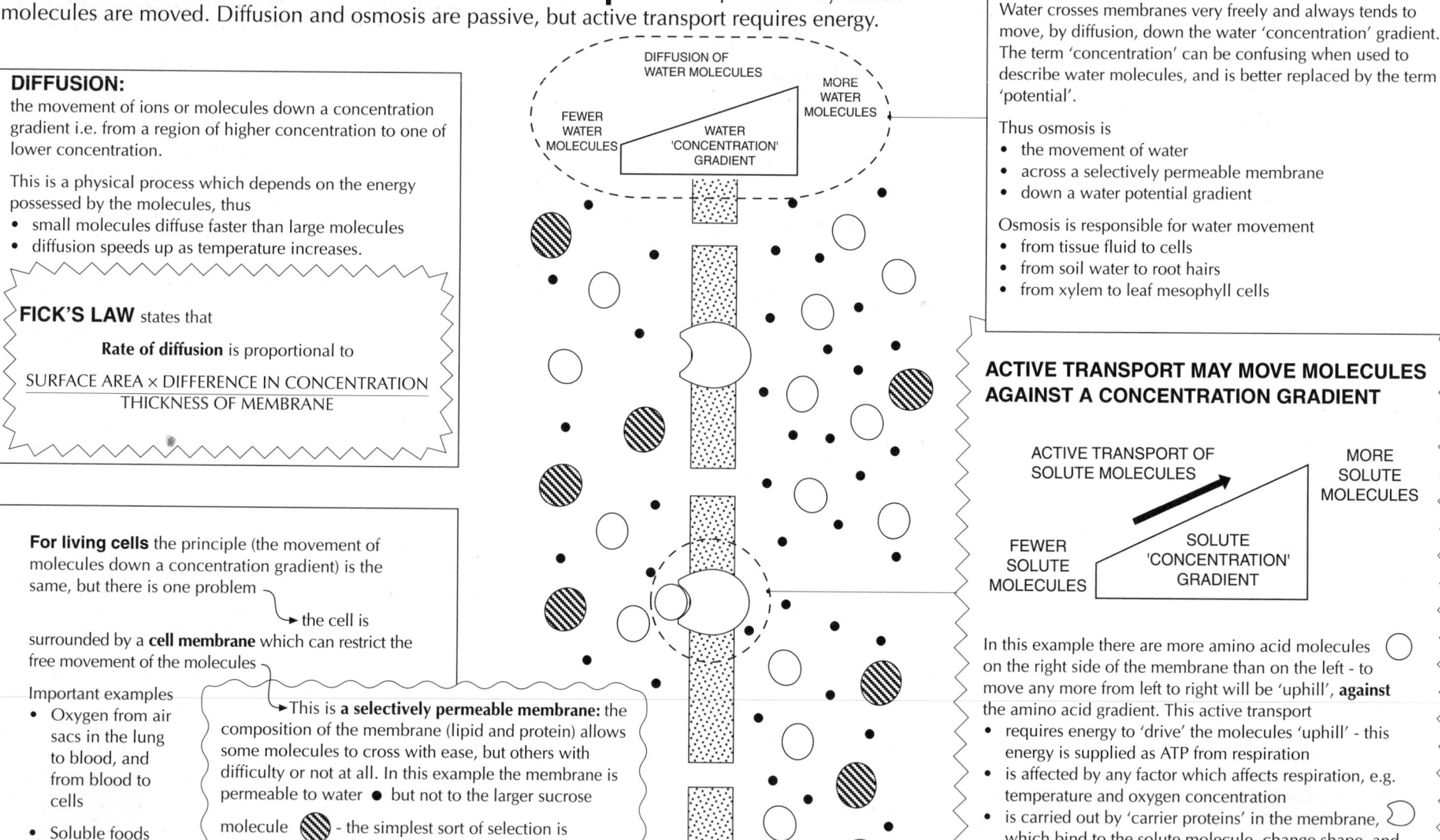

DIFFUSION OF WATER MOLECULES

FEWER WATER MOLECULES

WATER 'CONCENTRATION' GRADIENT

MORE WATER MOLECULES

OSMOSIS IS THE DIFFUSION OF WATER

Water crosses membranes very freely and always tends to move, by diffusion, down the water 'concentration' gradient. The term 'concentration' can be confusing when used to describe water molecules, and is better replaced by the term 'potential'.

Thus osmosis is
- the movement of water
- across a selectively permeable membrane
- down a water potential gradient

Osmosis is responsible for water movement
- from tissue fluid to cells
- from soil water to root hairs
- from xylem to leaf mesophyll cells

ACTIVE TRANSPORT MAY MOVE MOLECULES AGAINST A CONCENTRATION GRADIENT

ACTIVE TRANSPORT OF SOLUTE MOLECULES

MORE SOLUTE MOLECULES

FEWER SOLUTE MOLECULES

SOLUTE 'CONCENTRATION' GRADIENT

In this example there are more amino acid molecules on the right side of the membrane than on the left - to move any more from left to right will be 'uphill', **against** the amino acid gradient. This active transport
- requires energy to 'drive' the molecules 'uphill' - this energy is supplied as ATP from respiration
- is affected by any factor which affects respiration, e.g. temperature and oxygen concentration
- is carried out by 'carrier proteins' in the membrane, which bind to the solute molecule, change shape, and carry the molecule across the membrane.

Important examples are
- uptake of mineral ions from soil by root hair cells
- movement of sodium ions to set up nerve impulses

Water potential

Water potential is a measure of the free kinetic energy of water in a system, or the tendency of water to leave a system. It is measured in units of pressure (kPa) and is given the symbol ψ ('psi').

For pure water the water potential is arbitrarily given the value 0: this is a reference point, rather like the redox potential system used in chemistry.

i.e. for pure water $\psi = 0$

In a solution the presence of molecules of solute prevents water molecules leaving. Thus

$$\psi_{\text{SOLUTION}} < 0$$

(In the solution the solute molecules 'hinder' the movement of the water molecules, thus the kinetic energy of the water molecule is reduced and ψ becomes negative.)

Water moves down a gradient of water potential, i.e. from a less negative (e.g. -500 kPa) to a more negative (e.g. -1000 kPa) water potential.

The advantages of the water potential nomenclature

1. the movement of water is considered from the 'system's' point of view, rather than from that of the environment;

2. comparison between different systems can be made, e.g. between the atmosphere, the air in the spaces of a leaf and the leaf mesophyll cells.

We should remember that:

Osmosis is the movement of water, through a partially permeable membrane, along a water potential gradient.

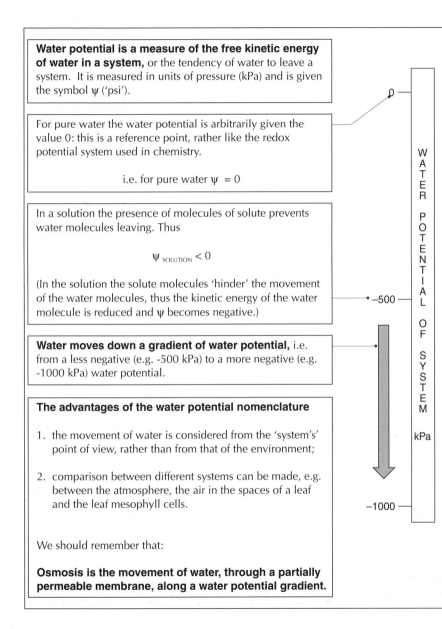

WATER POTENTIAL OF SYSTEM

0

−500

−1000

kPa

The presence of solutes in the cell sap makes ψ_s lower (more negative). The ψ for pure water = 0, so a plant cell placed in pure water will experience an inward movement of water, by osmosis, along a water potential gradient.

The cellulose cell wall is freely permeable to water and to solutes.

The plasma membrane is freely permeable to water but of limited permeability to solutes.

The entry of water will cause swelling of the protoplast (the cell contents inside the plasma membrane) causing a pressure (**turgor pressure**) to be exerted on the cell wall. Further expansion is resisted by a force **equal, but opposite in sign,** to the turgor pressure. This is called the **wall pressure** or **pressure potential** (ψ_p).

The pressure potential tends to resist the entry of water or to force water out of the cell.

ψ_{CELL}	=	ψ_s	+	ψ_p

Water potential of cell, i.e. the tendency of water to leave the cell: sometimes written as ψ_w, ψ_o or simply as ψ.

Effect of solute concentration on water potential – this is always negative in value.

Effect of wall pressure and turgor pressure on water potential – it represents the tendency for water to be forced out of the cell, and is either zero or positive in value.

Water movement between cells:

1. Calculate ψ for each cell from ψ_s and ψ_p.

2. Predict direction of water movement since water moves **down the gradient of water potential.**

Cell A

$\psi_s = -1300$ kPa
$\psi_p = -500$ kPa

$\psi = -1300 + 500$
$= -800$ kPa

Cell B

$\psi_s = -1900$ kPa
$\psi_p = 700$ kPa

$\psi = -1900 + 700$
$= -1200$ kPa

Since -800 is a higher number than -1200 **water will move by osmosis down a water potential gradient from A → B.**

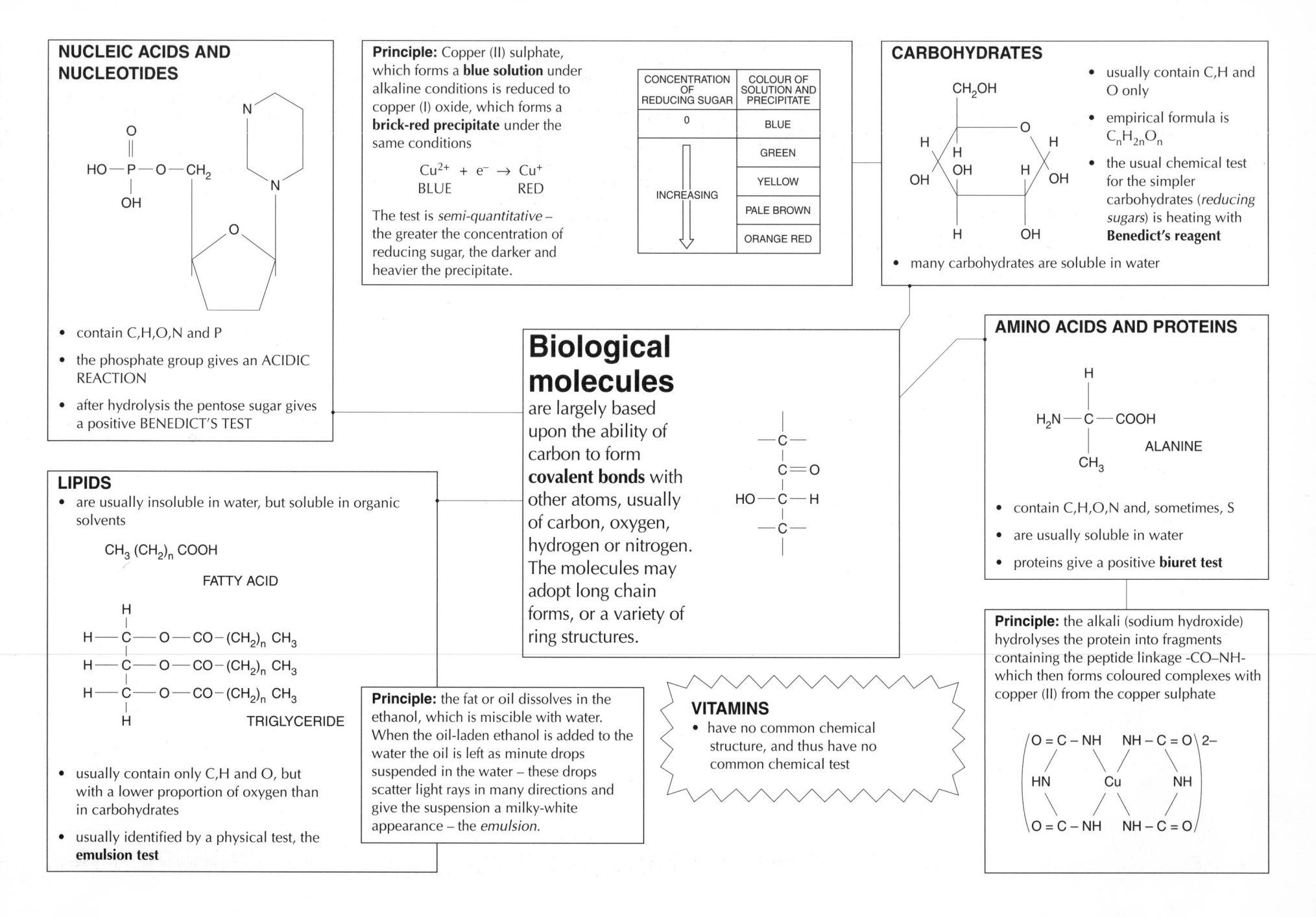

NUCLEIC ACIDS AND NUCLEOTIDES

- contain C,H,O,N and P
- the phosphate group gives an ACIDIC REACTION
- after hydrolysis the pentose sugar gives a positive BENEDICT'S TEST

Principle: Copper (II) sulphate, which forms a **blue solution** under alkaline conditions is reduced to copper (I) oxide, which forms a **brick-red precipitate** under the same conditions

$$Cu^{2+} + e^- \rightarrow Cu^+$$
BLUE   RED

The test is *semi-quantitative* – the greater the concentration of reducing sugar, the darker and heavier the precipitate.

CONCENTRATION OF REDUCING SUGAR	COLOUR OF SOLUTION AND PRECIPITATE
0	BLUE
	GREEN
INCREASING	YELLOW
	PALE BROWN
	ORANGE RED

CARBOHYDRATES

- usually contain C,H and O only
- empirical formula is $C_nH_{2n}O_n$
- the usual chemical test for the simpler carbohydrates (*reducing sugars*) is heating with **Benedict's reagent**
- many carbohydrates are soluble in water

Biological molecules

are largely based upon the ability of carbon to form **covalent bonds** with other atoms, usually of carbon, oxygen, hydrogen or nitrogen. The molecules may adopt long chain forms, or a variety of ring structures.

AMINO ACIDS AND PROTEINS

ALANINE

- contain C,H,O,N and, sometimes, S
- are usually soluble in water
- proteins give a positive **biuret test**

Principle: the alkali (sodium hydroxide) hydrolyses the protein into fragments containing the peptide linkage -CO–NH- which then forms coloured complexes with copper (II) from the copper sulphate

LIPIDS

- are usually insoluble in water, but soluble in organic solvents

$$CH_3 (CH_2)_n COOH$$

FATTY ACID

TRIGLYCERIDE

- usually contain only C,H and O, but with a lower proportion of oxygen than in carbohydrates
- usually identified by a physical test, the **emulsion test**

Principle: the fat or oil dissolves in the ethanol, which is miscible with water. When the oil-laden ethanol is added to the water the oil is left as minute drops suspended in the water – these drops scatter light rays in many directions and give the suspension a milky-white appearance – the *emulsion*.

VITAMINS

- have no common chemical structure, and thus have no common chemical test

Polysaccharides are polymers formed by glycosidic bonding of monosaccharide subunits

Starch is a mixture of two polymers of α-glucose: *amylose* typically contains about 300 glucose units joined by α 1,4 *glycosidic bonds*

The bulky –CH_2OH side chains cause the molecule to adopt a helical shape (excellent for packing many subunits into a limited space).

α-glucose molecules

Amylose helix (6 glucose units in each turn)

Amylopectin is a branched chain, containing up to 1500 glucose subunits, in which α 1,4 chains are cross-linked by α 1,6 *glycosidic bonds.*

α 1,6 GLYCOSIDIC BOND

Because there are so few 'ends' within the starch molecule there are few points to begin hydrolysis by the enzyme *amylase*. Starch is therefore an excellent long-term *storage compound.*

α-GLUCOSE

β-GLUCOSE

Glycogen is an α-glucose polymer, very similar to amylopectin but with very many more cross-links and shorter α 1,4 chains. This is appropriate to animal cells which may need to hydrolyse food reserves more rapidly than plant cells would do.

Cellulose is a polymer of glucose linked by β 1,4 *glycosidic bonds.* The β-conformation inverts successive monosaccharide units so that a straight chain polymer is formed.

β 1,4 GLYCOSIDIC BONDS

The parallel polysaccharide chains are then cross-linked by *hydrogen bonds.*

Hydrogen bonds

This cross-linking prevents access by water, so that cellulose is very resistant to hydrolysis and is therefore an excellent *structural molecule* (cellulose cell walls): ideal in plants which can readily synthesize excess carbohydrate.

Subunits are joined by *condensation* (**removal** of the elements of water) and separated by *hydrolysis* (bond breakage **using** the elements of water)

condensation

hydrolysis

+ H–OH

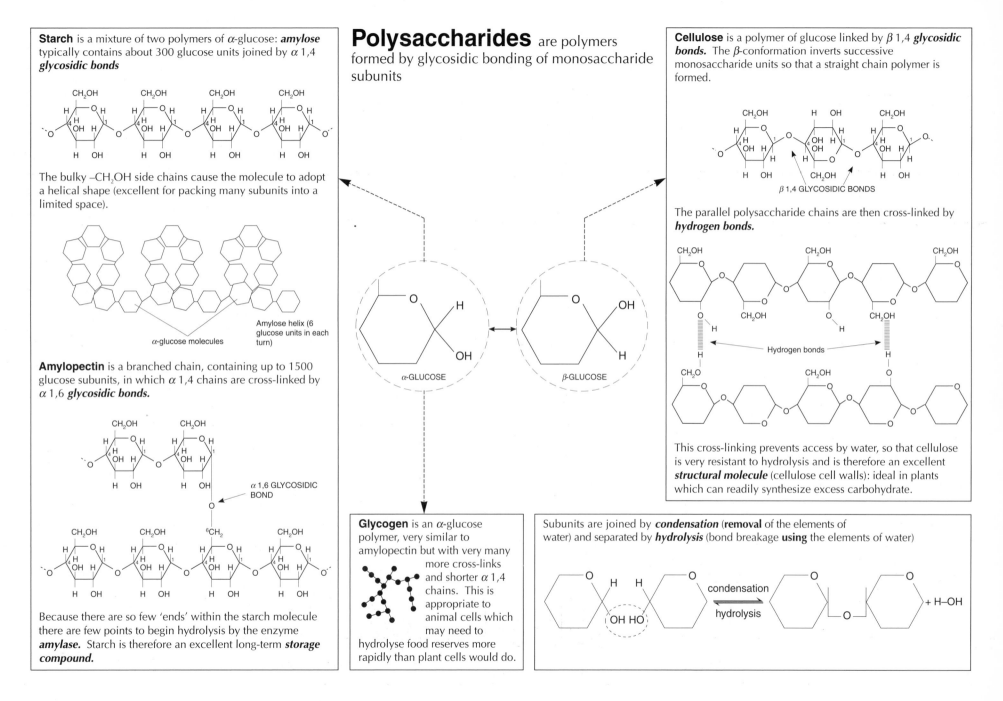

Functions of soluble carbohydrates include transport, protection, recognition and energy release.

Sugar derivatives include *sugar alcohols*, e.g. glycerol, *sugar acids*, e.g. ascorbic acid, and *mucopolysaccharides*, which are important components of connective tissues, synovial fluid, cartilage and bone. Heparin (anticoagulant in blood) is derived from mucopolysaccharides and has a protective function.

VAMPIRE BATS LIKE IT RUNNY!

Oligosaccharides are short (often 6-12 units) condensation products which combine with protein (*glycoprotein*) or lipid (*glycolipid*) and form the outer coat (*glycocalyx*) of animal cells. They are important in *cell-cell recognition* and the *immune response*.

INVADER

Glucose is the most common substrate for respiration (energy release).
Fructose is a constituent of seminal fluid. The dietary source is fruits, about 1 in 20 000 individuals suffers from *fructose intolerance* which may lead to *renal* and *liver damage*, or to *hypoglycaemia*.

Glucose and fructose are both *monosaccharides* (single sugar units) with the *empirical* formula $C_nH_{2n}O_n$. They each have *six carbon atoms* and are thus called *hexoses* (*pentoses* have 5 carbon atoms and *trioses* have 3).
Glucose and *fructose* are isomers of $C_6H_{12}O_6$.

α-GLUCOSE   α-FRUCTOSE

In naturally occurring **disaccharides** monosaccharide rings are joined together by *glycosidic bonds*.

This most usually occurs between *aldehyde* or *keto group* (i.e. the reducing group) of one monosaccharide and an *hydroxyl group* of another monosaccharide,

e.g. *lactose*

GALACTOSE → GLUCOSE

LACTOSE IS A REDUCING DISACCHARIDE

Reducing group of galactose   Hydroxyl group on C_4 of glucose   Reducing group of glucose = carbonyl group (C=O)

(*Maltose* is a reducing disaccharide formed from two molecules of α-glucose.)

or, more rarely, between *reducing groups of adjacent monosaccharides*,

e.g. *sucrose*

GLUCOSE FRUCTOSE

SUCROSE IS A NON-REDUCING DISACCHARIDE

Reducing groups are joined

Sucrose (*glucose-fructose*) is the main transport compound in plants. Commonly extracted from sugar cane and sugar beet and used as a sweetener.

TATE & LYLE

Lactose (*glucose-galactose*) is the carbohydrate source for suckling mammals - milk is about 5% lactose.

SEMI SKIM

Lactose intolerance occurs in many adults. This results from a deficiency in *lactase* so that dietary lactose accumulates in the lumen of the small intestine. This lowers the water potential of the gut contents causing an influx of fluid into the small intestine – this results in abdomenal distension, nausea, pain and diarrhoea. The condition is much more common in adult populations for whom milk is an unusual or uncommon food.

Cyclic AMP is an important intracellular messenger.

ADENINE

Ribose and **deoxyribose** are constituents of *nucleotides*

ORGANIC BASE

which are the subunits of *nucleic acids* (e.g. DNA).

CHAIN OF NUCLEOTIDES

Other important roles are in the *electron carriers* NAD, FAD and NADP and as the 'energy currency', Adenosine triphosphate.

ATP

ADENINE

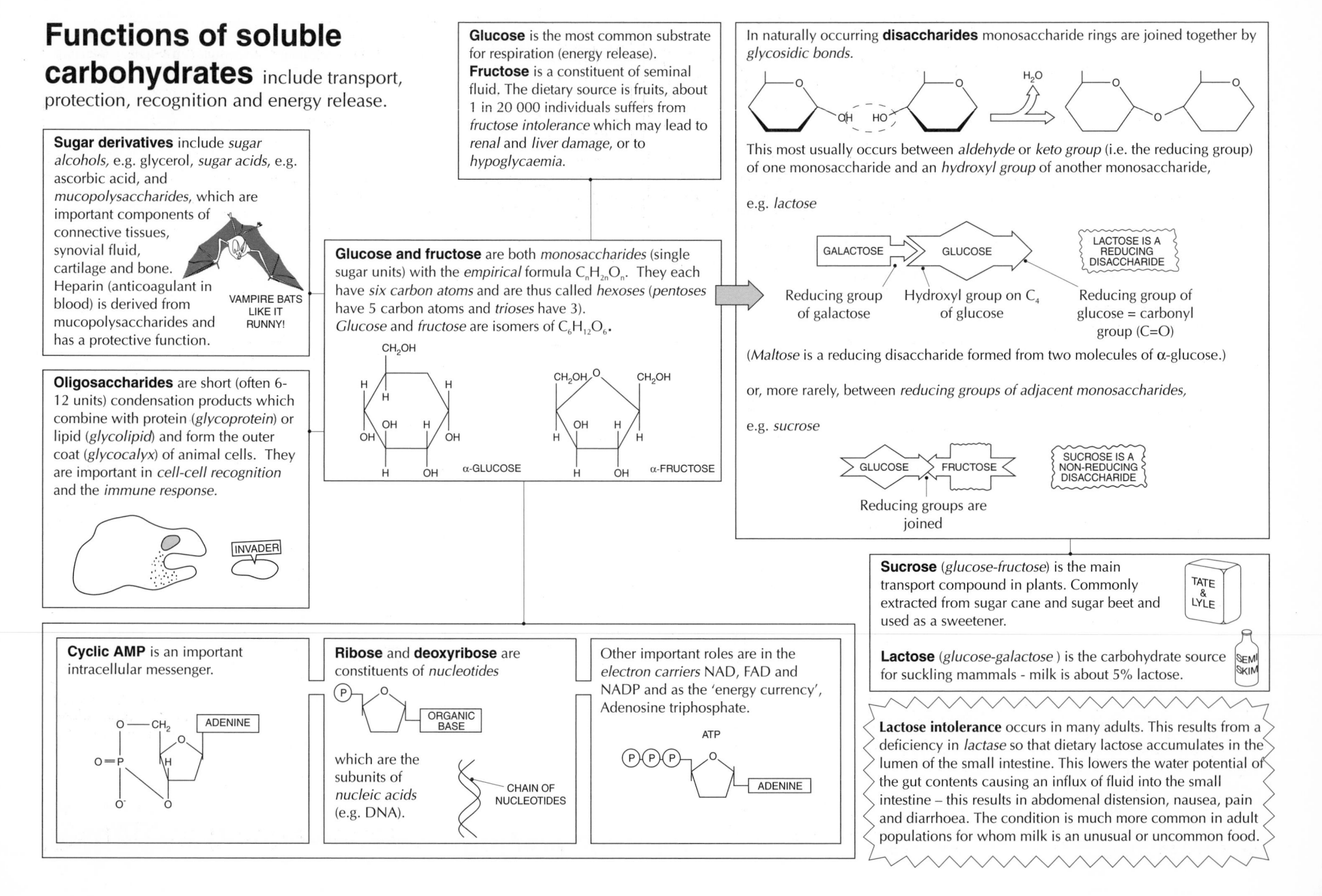

Testing for biochemicals

Aqueous solution, suspension or extract of test substance

Remove a 2 cm³ sample

Add a drop of Iodine solution (iodine in potassium iodide solution)

Positive result (blue-black coloration)
= **starch present**

STARCH   IODINE

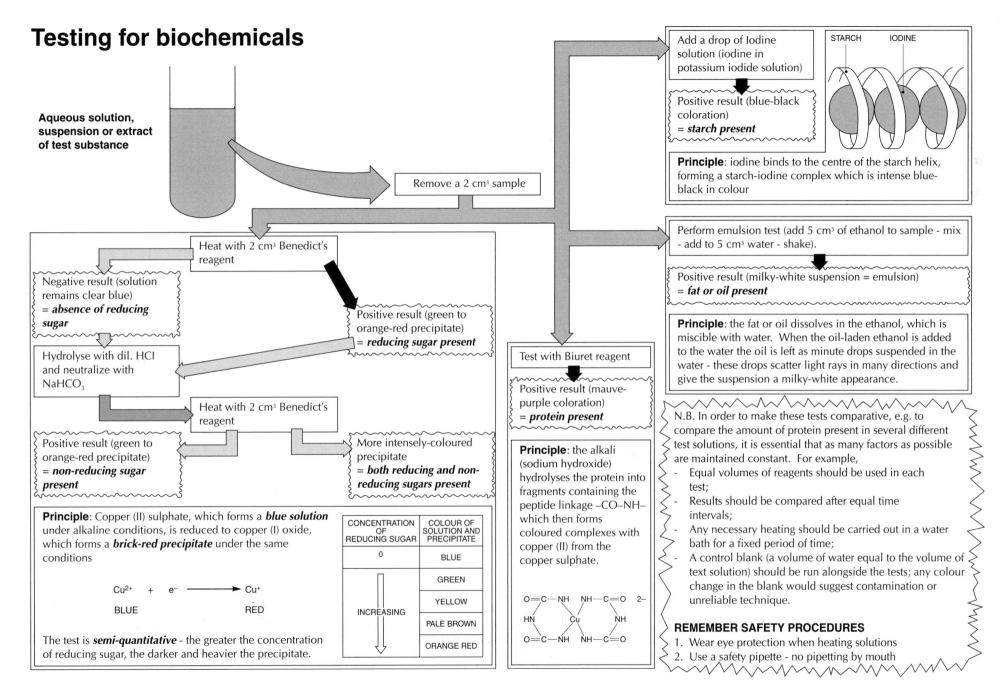

Principle: iodine binds to the centre of the starch helix, forming a starch-iodine complex which is intense blue-black in colour

Heat with 2 cm³ Benedict's reagent

Negative result (solution remains clear blue)
= **absence of reducing sugar**

Hydrolyse with dil. HCl and neutralize with NaHCO₃

Positive result (green to orange-red precipitate)
= **reducing sugar present**

Heat with 2 cm³ Benedict's reagent

Positive result (green to orange-red precipitate)
= **non-reducing sugar present**

More intensely-coloured precipitate
= **both reducing and non-reducing sugars present**

Test with Biuret reagent

Positive result (mauve-purple coloration)
= **protein present**

Perform emulsion test (add 5 cm³ of ethanol to sample - mix - add to 5 cm³ water - shake).

Positive result (milky-white suspension = emulsion)
= **fat or oil present**

Principle: the fat or oil dissolves in the ethanol, which is miscible with water. When the oil-laden ethanol is added to the water the oil is left as minute drops suspended in the water - these drops scatter light rays in many directions and give the suspension a milky-white appearance.

Principle: Copper (II) sulphate, which forms a **blue solution** under alkaline conditions, is reduced to copper (I) oxide, which forms a **brick-red precipitate** under the same conditions

$$Cu^{2+} + e^- \longrightarrow Cu^+$$

BLUE   RED

The test is **semi-quantitative** - the greater the concentration of reducing sugar, the darker and heavier the precipitate.

CONCENTRATION OF REDUCING SUGAR	COLOUR OF SOLUTION AND PRECIPITATE
0	BLUE
	GREEN
INCREASING	YELLOW
	PALE BROWN
	ORANGE RED

Principle: the alkali (sodium hydroxide) hydrolyses the protein into fragments containing the peptide linkage –CO–NH– which then forms coloured complexes with copper (II) from the copper sulphate.

$$O=C-NH \quad NH-C=O \quad 2-$$
$$HN \quad Cu \quad NH$$
$$O=C-NH \quad NH-C=O$$

N.B. In order to make these tests comparative, e.g. to compare the amount of protein present in several different test solutions, it is essential that as many factors as possible are maintained constant. For example,
- Equal volumes of reagents should be used in each test;
- Results should be compared after equal time intervals;
- Any necessary heating should be carried out in a water bath for a fixed period of time;
- A control blank (a volume of water equal to the volume of text solution) should be run alongside the tests; any colour change in the blank would suggest contamination or unreliable technique.

REMEMBER SAFETY PROCEDURES
1. Wear eye protection when heating solutions
2. Use a safety pipette - no pipetting by mouth

Levels of protein structure

There are twenty common amino acids, each of which is ingested as dietary protein or is formed by transamination in liver cells. All amino acids have the formula,

$$H_2N - C - COOH$$

where R represents the side chain which distinguishes one amino acid from another.

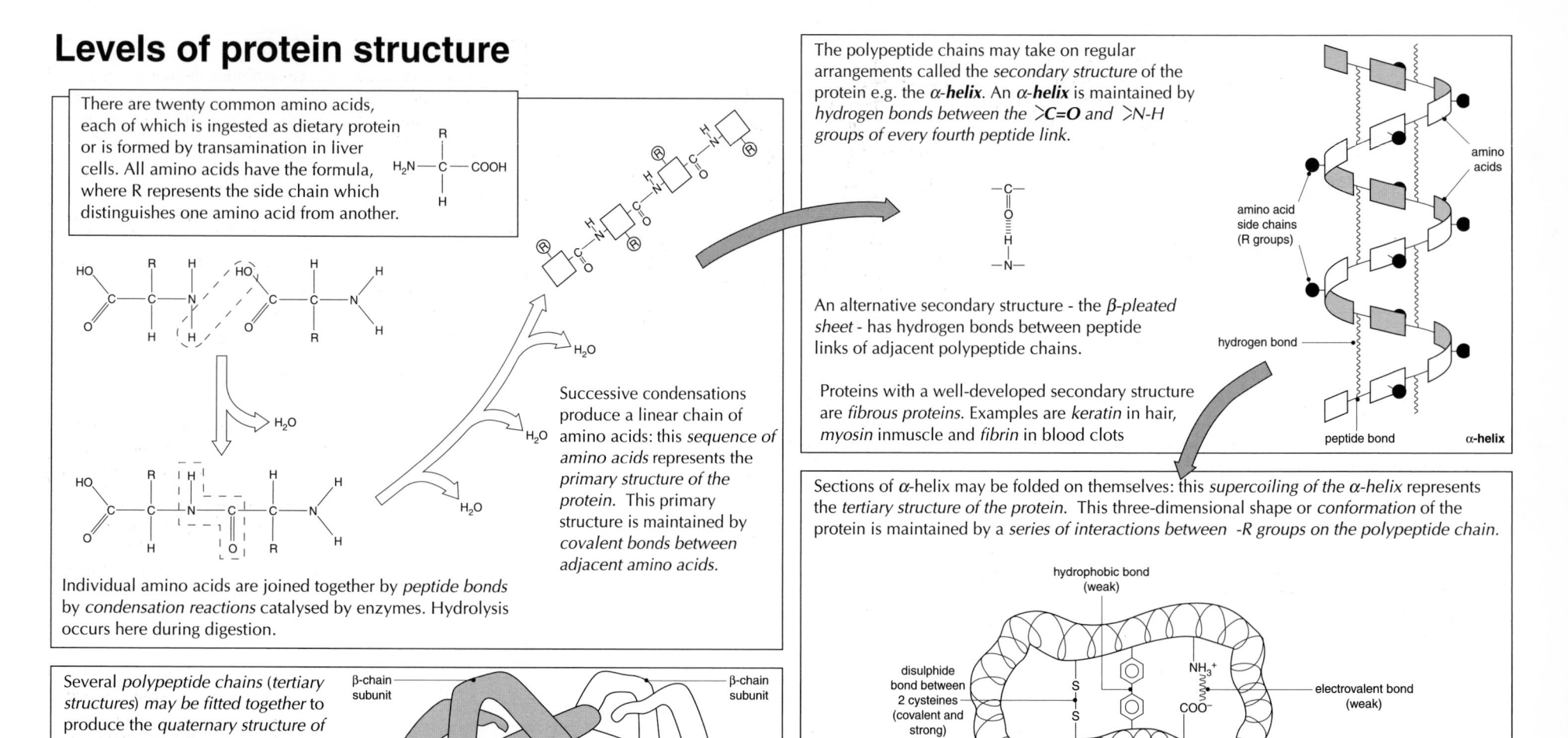

Successive condensations produce a linear chain of amino acids: this *sequence of amino acids* represents the *primary structure of the protein*. This primary structure is maintained by *covalent bonds between adjacent amino acids.*

Individual amino acids are joined together by *peptide bonds* by *condensation reactions* catalysed by enzymes. Hydrolysis occurs here during digestion.

The polypeptide chains may take on regular arrangements called the *secondary structure* of the protein e.g. the *α-helix*. An *α-helix* is maintained by *hydrogen bonds between the* $>C=O$ *and* $>N-H$ *groups of every fourth peptide link.*

An alternative secondary structure - the *β-pleated sheet* - has hydrogen bonds between peptide links of adjacent polypeptide chains.

Proteins with a well-developed secondary structure are *fibrous proteins*. Examples are *keratin* in hair, *myosin* in muscle and *fibrin* in blood clots

amino acids

amino acid side chains (R groups)

hydrogen bond

peptide bond   **α-helix**

Sections of α-helix may be folded on themselves: this *supercoiling of the α-helix* represents the *tertiary structure of the protein*. This three-dimensional shape or *conformation* of the protein is maintained by a *series of interactions between -R groups on the polypeptide chain.*

hydrophobic bond (weak)

disulphide bond between 2 cysteines (covalent and strong)

electrovalent bond (weak)

NH_3^+

COO^-

polypeptide α-helix held together by peptide linkages and hydrogen bonds

hydrogen bond (weak)

These interactions are very weak so that the conformation of such globular proteins can be easily altered by local physical changes - these alterations are reversible and are essential for the biological function of these molecules. Proteins with a well-developed tertiary structure are *globular proteins*. Examples are *enzymes, receptor molecules* in membranes and *serum albumin*.

Several *polypeptide chains* (*tertiary structures*) *may be fitted together* to produce the *quaternary structure of the protein*. The stability of the quaternary structure is maintained by *weak interactions between -R groups of adjacent polypeptide chains* and by *Van der Waal's forces between subunits.*

The relative movement of the polypeptide chains may be critical to the function of the protein. The oxygen transporter *haemoglobin* is an example of a protein with a quaternary structure

β-chain subunit

β-chain subunit

α-chain subunit

haem groups (total of 4 in complete haemoglobin molecule)

section of α-helix within α-chain subunit

Lipid structure and function

TRUE LIPIDS are esters of fatty acids and alcohols, formed by condensation reactions. Many of their properties result from their insolubility in water.

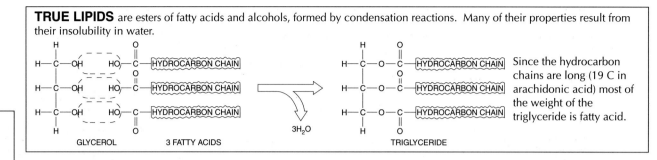

GLYCEROL   3 FATTY ACIDS $3H_2O$ TRIGLYCERIDE

Since the hydrocarbon chains are long (19 C in arachidonic acid) most of the weight of the triglyceride is fatty acid.

Water-repellent properties: oily secretions of the sebaceous glands help to waterproof the fur and skin. The preen gland of birds produces a secretion which performs a similar function on the feathers.

Cell membranes: phospholipids (phosphatides) are found in all cell membranes. These molecules have a polar 'phosphate-base' group substituted for one of the fatty acids in a triglyceride.

This part of the molecule

This part of the molecule is very *soluble* in water

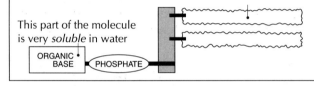

ORGANIC BASE   PHOSPHATE

Electrical insulation: myelin is secreted by Schwann cells and insulates some neurones in such a way that impulse transmission is made much more rapid.

Hormones: an important group of hormones, including cortisone, testosterone and oestrogen, are *steroids*. Steroids are not true esters but have the same solubility properties as them.

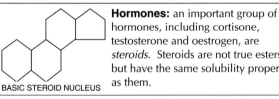

BASIC STEROID NUCLEUS

Physical protection: the shock-absorbing ability of subcutaneous fat stores protects delicate organs such as the kidneys from mechanical damage.

Thermal insulation: fats conduct heat very poorly - subcutaneous fat stores help heat retention in endothermic animals. Incompressible blubber is an important insulator in diving mammals.

FATS and oils are typical triglycerides which differ chemically in the nature of their hydrocarbon chains – these chains may be *saturated* ($[-CH_2.CH_2-]_n$) or partially *unsaturated* (contain some –C=C–bonds).

FATS have a high proportion of *saturated hydrocarbon chains*, and are *solid at room temperature*, whereas **oils** have a high proportion of *unsaturated hydrocarbon chains*, and are *liquid at room temperature*.

Blocked sebaceous glands may cause pimples or blackheads!

Nutrition: both bile acids and vitamin D (involved in fat digestion and Ca^{2+} absorption respectively) are manufactured from steroids.

Catalysis by enzymes

An important step in enzyme catalysis is substrate binding to the active sites.

Enzymes form **enzyme-substrate complexes** which reduce the activation energy for reactions which they catalyse.

Consider the reaction: SUBSTRATE (S) ⇌ PRODUCT (P)

which can be illustrated by a **reaction profile.**

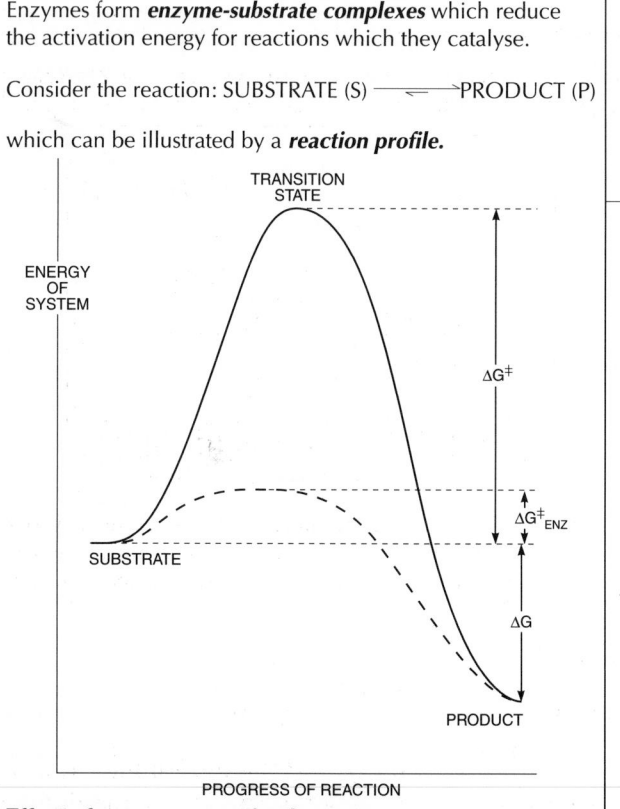

ENERGY OF SYSTEM

TRANSITION STATE

$\Delta G^{\ddagger}$

$\Delta G^{\ddagger}_{ENZ}$

SUBSTRATE

ΔG

PRODUCT

PROGRESS OF REACTION

Effect of enzyme on activation energy

For a reaction S⇌P the rate of the forward reaction depends on temperature and activation energy (difference in free energy between substrate and transition state, $\Delta G^{\ddagger}$). The reaction rate is proportional to the number of molecules which have an energy $\geq \Delta G^{\ddagger}$. **Enzymes act as catalysts by providing alternative reaction pathways in which $\Delta G^{\ddagger}$ is lower than it otherwise would be.** Heat cannot be used by cells to increase rates of reaction because of possible denaturation.

$$E + S \rightleftharpoons E{-}S \longrightarrow E + P$$

ENZYME-SUBSTRATE COMPLEX

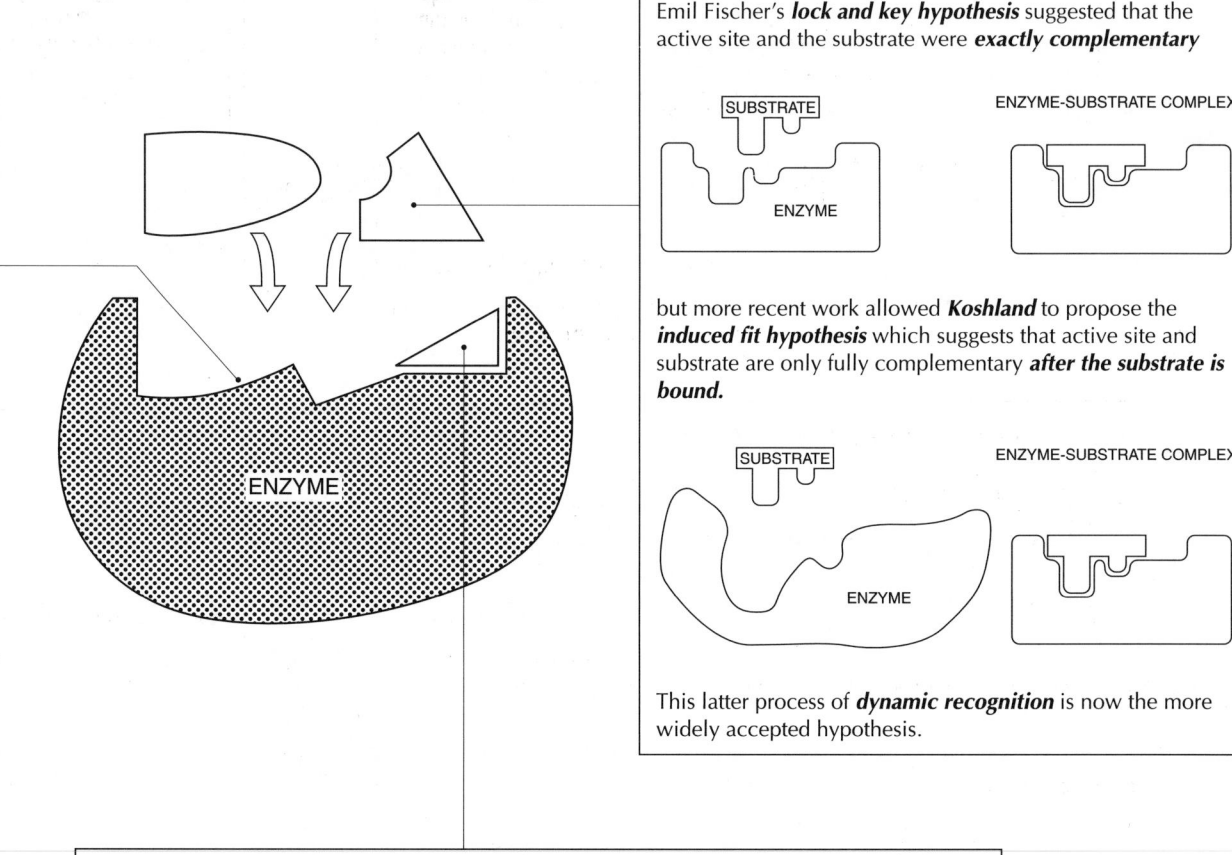

ENZYME

Stereospecificity: relationship of substrate(s) to active site

Emil Fischer's **lock and key hypothesis** suggested that the active site and the substrate were **exactly complementary**

SUBSTRATE

ENZYME-SUBSTRATE COMPLEX

ENZYME

but more recent work allowed **Koshland** to propose the **induced fit hypothesis** which suggests that active site and substrate are only fully complementary **after the substrate is bound.**

SUBSTRATE

ENZYME-SUBSTRATE COMPLEX

ENZYME

This latter process of **dynamic recognition** is now the more widely accepted hypothesis.

Cofactors are essential for enzyme activity

Some, such as Zn^{2+} or Mg^{2+}, or porphyrin groups such as the **haem** in catalase, may form part of the active site and cannot easily be separated from the enzyme protein: these are commonly called **prosthetic groups.**

Some, such as NAD (nicotinamide adenine dinucleotide), bind temporarily to the active site and actually take part in the reaction.

e.g. lactate + NAD $\underset{\text{DEHYDROGENASE}}{\overset{\text{LACTATE}}{\rightleftharpoons}}$ pyruvate + $NADH_2$

Such **coenzymes** shuttle between one enzyme system and another - most are formed from dietary components called **vitamins** (e.g. NAD is formed from niacin, one of the B vitamin complex).

Naming and classifying enzymes

Originally the digestive enzymes of the stomach and the small intestine were given names ending in -*in*: thus peps-*in* and chymotryps-*in*. Now a more widely used way of naming enzymes has been to add the suffix -*ase* to the reaction catalysed. Thus

    a hydrolase catalyses a hydrolysis

    a dehydrogenase catalyses the removal of hydrogen

A complete 'trivial' (everyday) name for an enzyme would also include the substrate for the enzyme. Thus

CREATINE	PHOSPHOKINASE
Substrate	Type of reaction

A rigid numerical classification for all enzymes has now been devised and accepted internationally. It recognises that there are only six basic reactions catalysed by enzymes and gives an Enzyme Commission Number to every enzyme.

BIOCHEMICAL OLYMPICS

Phew! My 1.1.1.27 has had it!

1. Oxidoreductases

2. Transferases

3. Hydrolases

4. Lyases

5. Isomerases

6. Ligases

Within each group further subdivisions are made so that each enzyme is unambiguously identified by a set of four numbers. For example

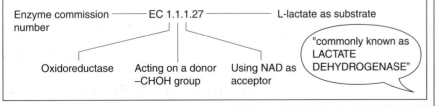

Enzyme commission number ———— EC 1.1.1.27 ———— L-lactate as substrate

Oxidoreductase   Acting on a donor –CHOH group   Using NAD as acceptor

"commonly known as LACTATE DEHYDROGENASE"

catalyse **redox** reactions i.e. the transfer of H and O atoms, or electrons from one substance to another

e.g. LACTATE DEHYDROGENASE

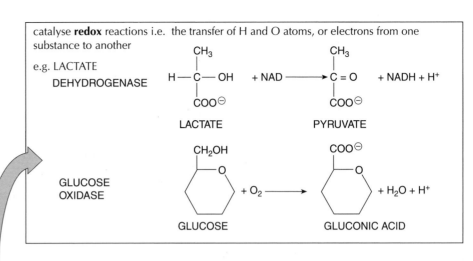

LACTATE   PYRUVATE

GLUCOSE OXIDASE

GLUCOSE   GLUCONIC ACID

catalyse **hydrolysis** reactions i.e. the breakdown of one molecule to two products by the addition of the elements of water

e.g. LIPASE

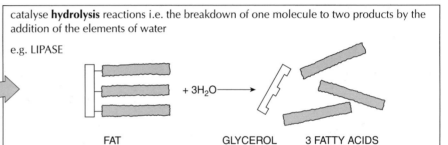

FAT   GLYCEROL   3 FATTY ACIDS

catalyse **synthetic** reactions i.e. the joining together of two molecules by the formation of new C-O, C-C, C-N or C-S bonds at the same time as ATP is broken down

e.g. AMINOACYL tRNA SYNTHETASE

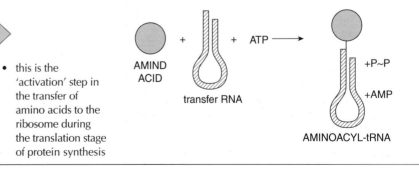

- this is the 'activation' step in the transfer of amino acids to the ribosome during the translation stage of protein synthesis

AMIND ACID   +   transfer RNA   + ATP   →   AMINOACYL-tRNA   +P~P   +AMP

Factors affecting enzyme activity exert their effects by altering the ease with which an enzyme-substrate complex is formed.

Any factor which alters the conformation (dependent on tertiary structure) of the enzyme will alter the shape of the active site, affect the frequency of enzyme-substrate complex formation and thus influence the rate of the enzyme-catalysed reaction.

EFFECT OF TEMPERATURE

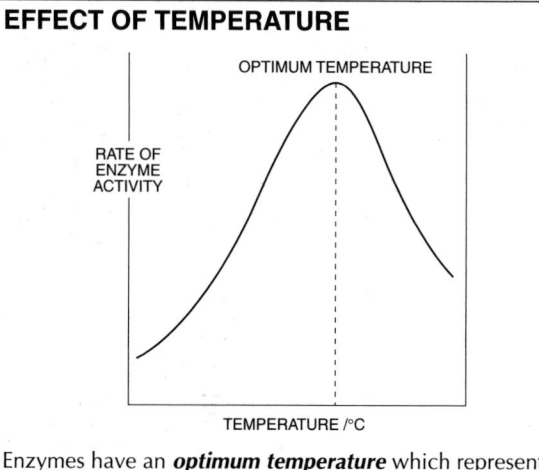

Enzymes have an **optimum temperature** which represents a compromise between **activation** due to increased rate of collision between E and S and **loss of activity** due to denaturation of E molecules and consequent distortion of the active site.

EFFECT OF pH

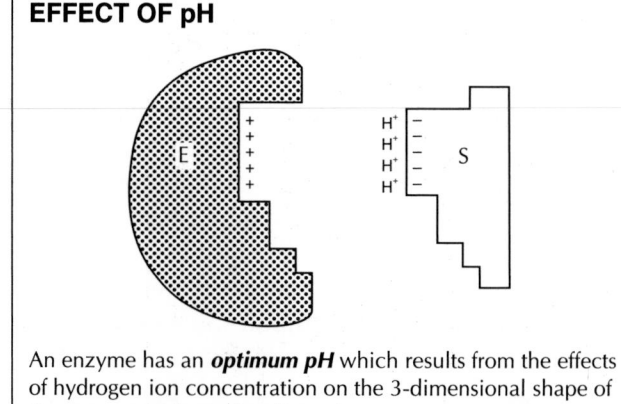

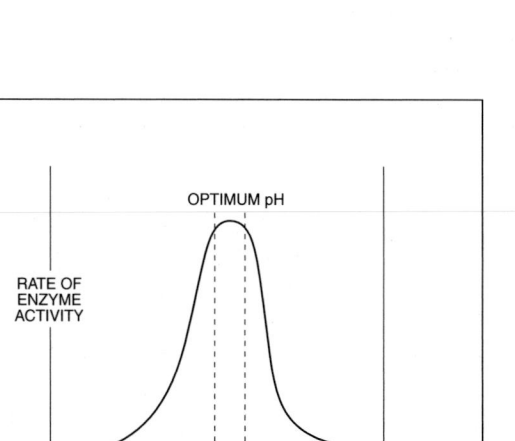

An enzyme has an **optimum pH** which results from the effects of hydrogen ion concentration on the 3-dimensional shape of the enzyme in the active site region.

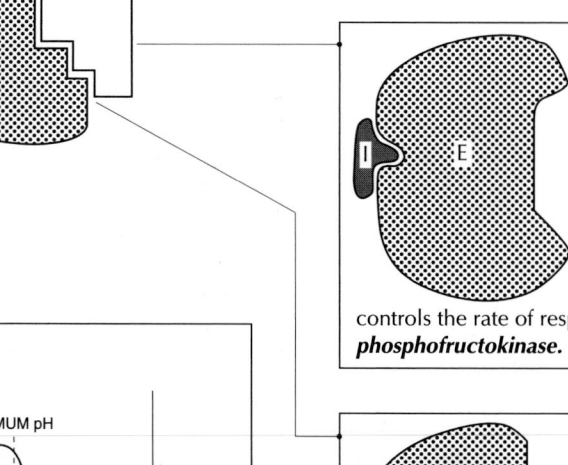

Competitive inhibitors compete for the active site with the normal substrate. These inhibitors therefore must have a similar structure to the natural substrate.
The success of the binding of I to the active site depends on the relative concentrations of I and S, and such inhibition is therefore **reversible by an increase in substrate concentration,** e.g. malonate competes with succinate for the active site on the enzyme **succinate dehydrogenase.**

Irreversible inhibition occurs if the enzyme-inhibitor binding is covalent and the distortion of the active site may be permanent, e.g. cyanide (CN⁻) binds irreversibly to the active site of the enzyme **cytochrome oxidase.**

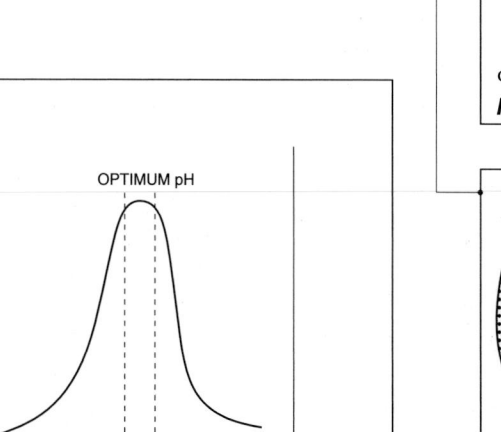

Non-competitive inhibitors reduce enzyme activity by distortion of enzyme conformation caused by binding to some site **other than the active site.** If the binding is non-covalent the inhibition may be **reversible if the inhibitor concentration is diminished.** Many such inhibitors are natural **allosteric regulators** of metabolism, e.g. ATP controls the rate of respiration by inhibition of the enzyme **phosphofructokinase.**

Activators may be necessary to complete the structural relationship between active site and substrate, e.g. chloride ions (Cl⁻) are required for activity of the enzyme **salivary amylase.**

There are also **allosteric activators** which enhance enzyme-substrate binding by alteration of enzyme conformation when binding to another ('**allosteric**') site on the enzyme.

Commercial applications of enzymes

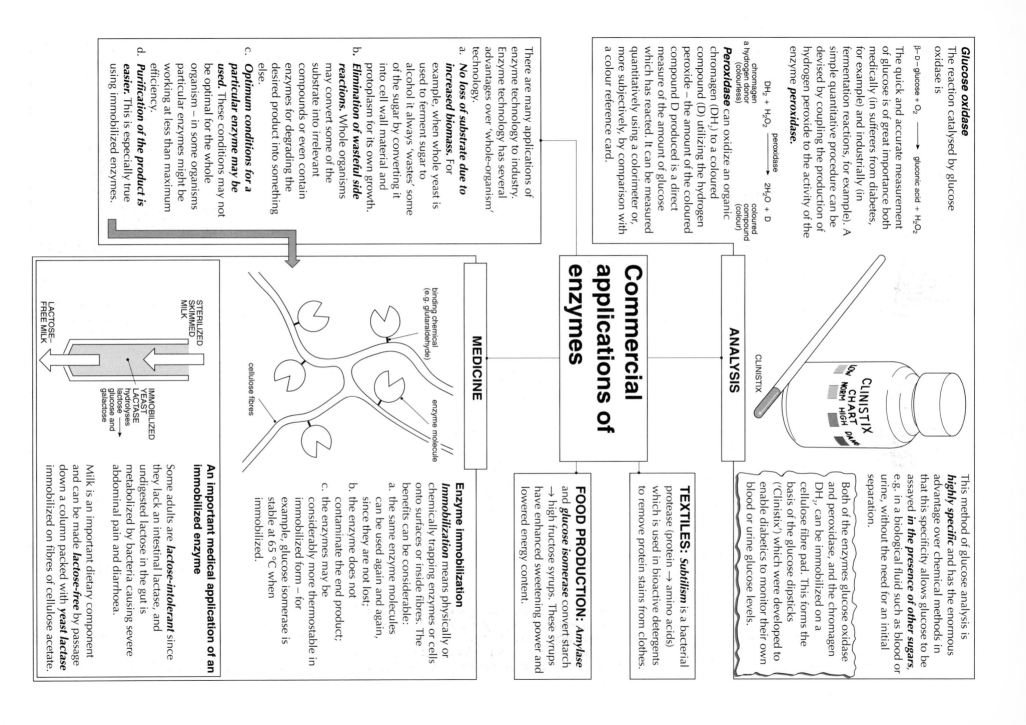

Glucose oxidase

The reaction catalysed by glucose oxidase is

$$\beta\text{-}D\text{-glucose} + O_2 \longrightarrow \text{gluconic acid} + H_2O_2$$

The quick and accurate measurement of glucose is of great importance both medically (in sufferers from diabetes, for example) and industrially (in fermentation reactions, for example). A simple quantitative procedure can be devised by coupling the production of hydrogen peroxide to the activity of the enzyme **peroxidase**.

Peroxidase can oxidize an organic chromagen (DH₂) to a coloured compound (D) utilizing the hydrogen peroxide – the amount of the coloured compound D produced is a direct measure of the amount of glucose which has reacted. It can be measured quantitatively using a colorimeter or, more subjectively, by comparison with a colour reference card.

$$DH_2 + H_2O_2 \xrightarrow{\text{peroxidase}} 2H_2O + D$$

chromagen (colourless) / a hydrogen donor (colourless) → coloured compound (colour)

ANALYSIS

This method of glucose analysis is **highly specific** and has the enormous advantage over chemical methods in that this specificity allows glucose to be assayed **in the presence of other sugars**, e.g. in a biological fluid such as blood or urine, without the need for an initial separation.

Both of the enzymes glucose oxidase and peroxidase, and the chromagen DH₂, can be immobilized on a cellulose fibre pad. This forms the basis of the glucose dipsticks ('Clinistix') which were developed to enable diabetics to monitor their own blood or urine glucose levels.

CLINISTIX — LOW CHART / NORM / HIGH / DANG

Enzyme technology

There are many applications of enzyme technology to industry. Enzyme technology has several advantages over 'whole-organism' technology.

a. **No loss of substrate due to increased biomass.** For example, when whole yeast is used to ferment sugar to alcohol it always 'wastes' some of the sugar by converting it into cell wall material and protoplasm for its own growth.

b. **Elimination of wasteful side reactions.** Whole organisms may convert some of the substrate into irrelevant compounds or even contain enzymes for degrading the desired product into something else.

c. **Optimum conditions for a particular enzyme may be used.** These conditions may not be optimal for the whole organism – in some organisms particular enzymes might be working at less than maximum efficiency.

d. **Purification of the product is easier.** This is especially true using immobilized enzymes.

MEDICINE

Enzyme immobilization

Immobilization means physically or chemically trapping enzymes or cells onto surfaces or inside fibres. The benefits can be considerable:

a. the same enzyme molecules can be used again and again, since they are not lost;

b. the enzyme does not contaminate the end product;

c. the enzymes may be considerably more thermostable in immobilized form – for example, glucose isomerase is stable at 65 °C when immobilized.

binding chemical (e.g. glutaraldehyde) — cellulose fibres — enzyme molecule

An important medical application of an immobilized enzyme

Some adults are **lactose-intolerant** since they lack an intestinal lactase, and undigested lactose in the gut is metabolized by bacteria causing severe abdominal pain and diarrhoea.

Milk is an important dietary component and can be made **lactose-free** by passage down a column packed with **yeast lactase** immobilized on fibres of cellulose acetate.

STERILIZED SKIMMED MILK → IMMOBILIZED YEAST LACTASE hydrolyses lactose → glucose and galactose → LACTOSE-FREE MILK

FOOD PRODUCTION

FOOD PRODUCTION: *Amylase* and *glucose isomerase* convert starch → high fructose syrups. These syrups have enhanced sweetening power and lowered energy content.

TEXTILES

TEXTILES: *Subtilism* is a bacterial protease (protein → amino acids) which is used in bioactive detergents to remove protein stains from clothes.

Respirometers and the measurement of respiratory quotient

PRINCIPLE: Carbon dioxide evolved during respiration is absorbed by potassium hydroxide solution. If the system is closed to the atmosphere a change in volume of the gas within the chamber must be due to the consumption of oxygen. The change in volume, i.e. the oxygen consumption, is measured as the movement of a drop of coloured fluid along a capillary tube.

Coloured fluid. Narrow bore capillary tube ensures maximum movement of fluid with any change in volume of gas in chamber.

Graduated scale against which movement of coloured fluid may be measured.

Gauze basket to hold respiring material must be porous to permit free exchange of gases.

Weighed amount of respiring material. The mass, m, of the respiring material should be known so that results can be made comparative.

Glass rod to keep respiring material out of direct contact with potassium hydroxide solution.

Spring clip. When open permits equilibration of contents of chamber with atmosphere.

Rubber stopper. Airtightness (which is essential) can be improved by smearing vaseline along seal between chamber and stopper.

Respiratory chamber has relatively low volume to ensure that volume changes due to respiratory activity are significant enough to be measured.

Filter paper wick to ensure maximum surface area of potassium hydroxide solution is available to contents of chamber.

Potassium hydroxide solution to absorb carbon dioxide evolved during respiration.

RESPIRATORY QUOTIENT (R.Q.)

$$= \frac{CO_2 \text{ evolved}}{O_2 \text{ consumed}} \text{ during respiration of a particular food}$$

Aerobic respiration of carbohydrate

$$C_6H_{12}O_6 + 6O_2 \rightarrow 6CO_2 + 6H_2O$$

Thus R.Q. $= \dfrac{6}{6} = 1.0$

For lipid R.Q. ≈ 0.7, for protein R.Q. ≈ 0.95

Thus the nature of the respiratory substrate can be determined by measurement of R.Q.

Mixed diet (ChO/lipid) R.Q. ≈ 0.85

Starvation (body protein reserves) R.Q. $\approx 0.9 - 1.0$

METHOD

1 The apparatus, containing a known mass of respiring material, is set up with the spring clip open to permit equilibration.

2 With the spring clip closed the time, t, in minutes, for the coloured fluid to move a distance l mm. along the capillary tube is noted.

   Oxygen Consumed $= l$ units

3 Remove KOH solution and note fluid movement in same time, t

   Oxygen Consumed – Carbon Dioxide Released $= m$ units

   Then R.Q. $= \dfrac{l - m}{l}$

Glycolysis generates ATP, reduced electron carriers and pyruvate.

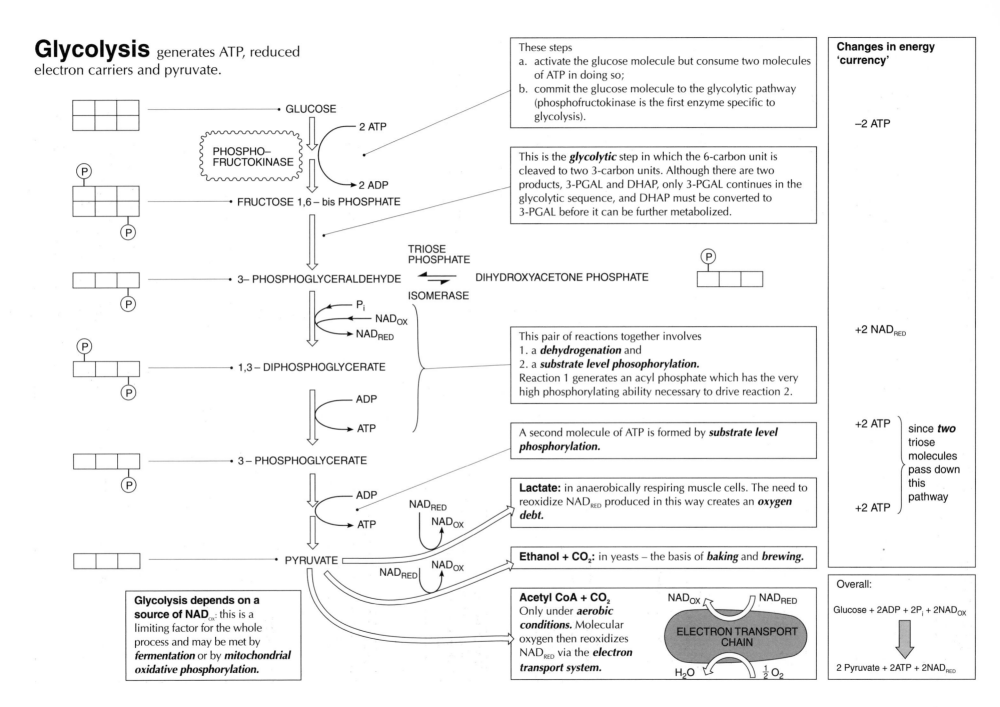

GLUCOSE

2 ATP

PHOSPHO–FRUCTOKINASE

2 ADP

FRUCTOSE 1,6 – bis PHOSPHATE

TRIOSE PHOSPHATE

3– PHOSPHOGLYCERALDEHYDE ⇌ DIHYDROXYACETONE PHOSPHATE

ISOMERASE

P_i
NAD_{OX}
NAD_{RED}

1,3 – DIPHOSPHOGLYCERATE

ADP

ATP

3 – PHOSPHOGLYCERATE

ADP

ATP

NAD_{RED}
NAD_{OX}

NAD_{RED}
NAD_{OX}

PYRUVATE

These steps
a. activate the glucose molecule but consume two molecules of ATP in doing so;
b. commit the glucose molecule to the glycolytic pathway (phosphofructokinase is the first enzyme specific to glycolysis).

This is the **glycolytic** step in which the 6-carbon unit is cleaved to two 3-carbon units. Although there are two products, 3-PGAL and DHAP, only 3-PGAL continues in the glycolytic sequence, and DHAP must be converted to 3-PGAL before it can be further metabolized.

This pair of reactions together involves
1. a **dehydrogenation** and
2. a **substrate level phosophorylation.**
Reaction 1 generates an acyl phosphate which has the very high phosphorylating ability necessary to drive reaction 2.

A second molecule of ATP is formed by **substrate level phosphorylation.**

Lactate: in anaerobically respiring muscle cells. The need to reoxidize NAD_{RED} produced in this way creates an **oxygen debt.**

Ethanol + CO₂: in yeasts – the basis of **baking** and **brewing.**

Acetyl CoA + CO₂
Only under **aerobic conditions.** Molecular oxygen then reoxidizes NAD_{RED} via the **electron transport system.**

NAD_{OX} NAD_{RED}

ELECTRON TRANSPORT CHAIN

H_2O $\frac{1}{2}O_2$

Glycolysis depends on a source of NAD_{OX}: this is a limiting factor for the whole process and may be met by **fermentation** or by **mitochondrial oxidative phosphorylation.**

Changes in energy 'currency'

−2 ATP

+2 NAD_{RED}

+2 ATP ⎤
       ⎥ since **two** triose molecules pass down this pathway
+2 ATP ⎦

Overall:

Glucose + 2ADP + 2P_i + 2NAD_{OX}

⬇

2 Pyruvate + 2ATP + 2NAD_{RED}

The TCA (Krebs) cycle is essential for energy release from, and interconversion of, a variety of nutrients.

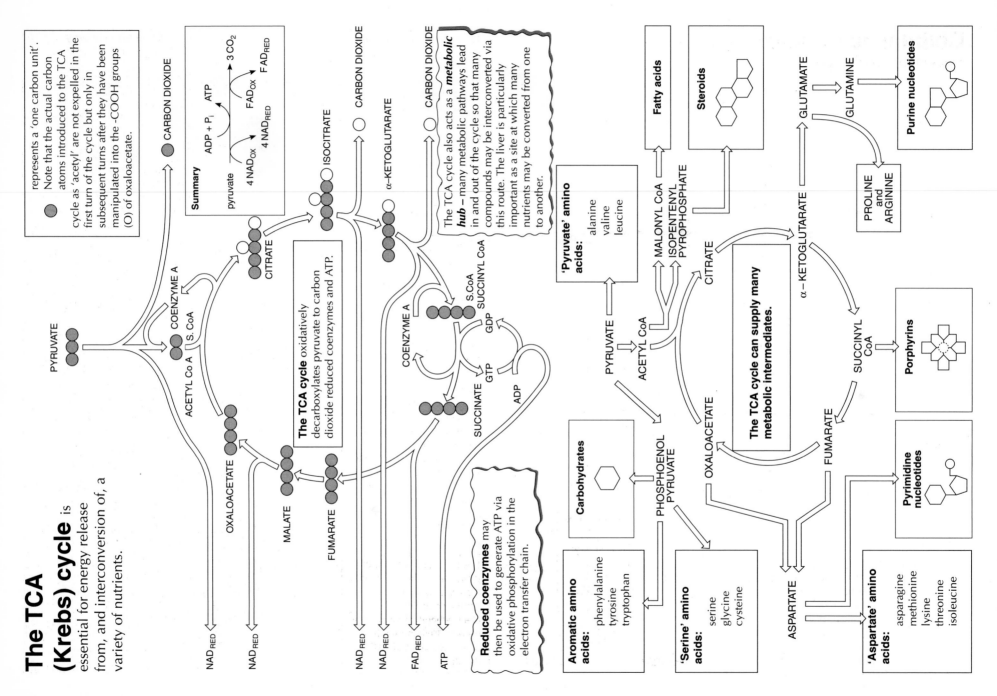

represents a 'one carbon unit'. Note that the actual carbon atoms introduced to the TCA cycle as 'acetyl' are not expelled in the first turn of the cycle but only in subsequent turns after they have been manipulated into the -COOH groups (O) of oxaloacetate.

Summary

$$pyruvate \xrightarrow{4\,NAD_{OX}} 4\,NAD_{RED}$$
$$ADP + P_i \rightarrow ATP$$
$$FAD_{OX} \rightarrow FAD_{RED}$$
$$\rightarrow 3\,CO_2$$

The TCA cycle oxidatively decarboxylates pyruvate to carbon dioxide reduced coenzymes and ATP.

Reduced coenzymes may then be used to generate ATP via oxidative phosphorylation in the electron transfer chain.

The TCA cycle also acts as a **metabolic hub** – many metabolic pathways lead in and out of the cycle so that many compounds may be interconverted via this route. The liver is particularly important as a site at which many nutrients may be converted from one to another.

The TCA cycle can supply many metabolic intermediates.

PYRUVATE
CARBON DIOXIDE
COENZYME A
ACETYL Co A   S. CoA
CITRATE
ISOCITRATE
CARBON DIOXIDE
α-KETOGLUTARATE
CARBON DIOXIDE
COENZYME A
S.CoA   SUCCINYL CoA
GDP   GTP
ADP
SUCCINATE
FUMARATE
MALATE
OXALOACETATE

NAD_RED
NAD_RED
NAD_RED
NAD_RED
FAD_RED
ATP

Carbohydrates

Aromatic amino acids: phenylalanine, tyrosine, tryptophan

'Serine' amino acids: serine, glycine, cysteine

PHOSPHOENOL PYRUVATE

PYRUVATE
ACETYL CoA

'Pyruvate' amino acids: alanine, valine, leucine

MALONYL CoA
ISOPENTENYL PYROPHOSPHATE

Fatty acids

Steroids

CITRATE
α – KETOGLUTARATE
GLUTAMATE
GLUTAMINE

Purine nucleotides

PROLINE AND ARGININE

OXALOACETATE
FUMARATE
SUCCINYL CoA

Porphyrins

Pyrimidine nucleotides

ASPARTATE

'Aspartate' amino acids: asparagine, methionine, lysine, threonine, isoleucine

Cellular respiration occurs in a series of localized stages.

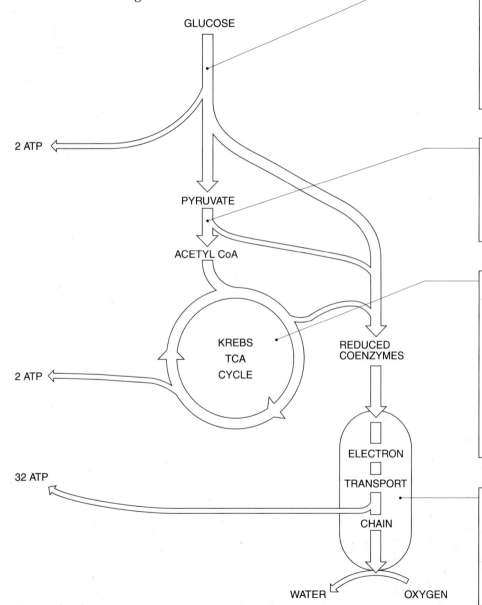

GLUCOSE

First stage: glycolysis is a series of about ten steps by which glucose is degraded ('lysed') to two molecules of pyruvate. Two molecules of ATP and two molecules of reduced coenzyme are generated. This stage can occur anaerobically.

2 ATP

PYRUVATE

Second stage: 'activation' of pyruvate to acetyl CoA 'drives' the pyruvate molecule towards the TCA cycle. Two molecules of reduced coenzyme are generated.

ACETYL CoA

KREBS TCA CYCLE

Third stage: the Krebs (tricarboxylic acid) cycle is a series of dehydrogenations, decarboxylations and isomerizations. One molecule of ATP and four molecules of reduced coenzyme are generated for *each turn of the cycle.* (The cycle is 'turned' twice for each glucose molecule.) Aerobic.

2 ATP

REDUCED COENZYMES

ELECTRON

TRANSPORT

32 ATP

CHAIN

WATER   OXYGEN

Fourth stage: oxidative phosphorylation in which a proton gradient is generated and its electrochemical potential is used to drive the synthesis of 32 molecules of ATP. Aerobic.

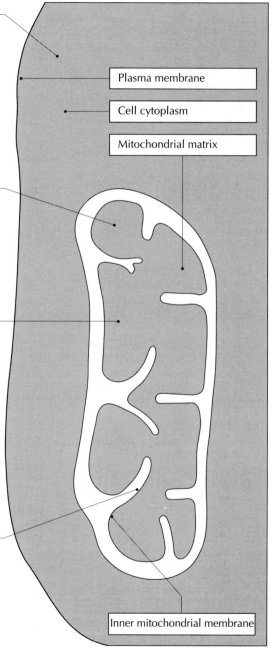

Plasma membrane

Cell cytoplasm

Mitochondrial matrix

Inner mitochondrial membrane

ATP: the energy currency of the cell

This part of the molecule contains anhydride bonds (O-P) which can be hydrolysed in reactions which are *exergonic* (energy-yielding) and can be coupled to *endergonic* (energy-demanding) reactions.

This part of the molecule acts like a 'handle' – it has a shape which can be recognized by highly specific enzymes.

ADENOSINE TRIPHOSPHATE

ADENOSINE

ADENINE

RIBOSE

ATP hydrolysis is favoured

$ATP^{4-} + H_2O \longrightarrow ADP^{3-} + P^{2-} + H^+ + 30.5$ kJ mol^{-1}

i.e. ATP has a strong tendency to transfer its terminal phosphoryl group, a reaction associated with the release of 30.5 kJ mol^{-1} of ATP, because

1. the repulsion between the four negative charges in ATP^{4-} is reduced when ATP is hydrolysed because two negative charges are removed with phosphate.

2. the H$^+$ ion which is released when ATP is hydrolysed reacts with OH$^-$ ions to form water – this is a highly favoured reaction.

3. the charge distribution on ADP + P is more stable than that on ATP.

Substrate level phosphorylation: A phosphate group is transferred from a phosphorylated compound to ADP.

e.g.

PHOSPHOENOL PYRUVATE

PYRUVATE

ADP

ATP

Muscle contraction: ATP hydrolysis changes the position of the myosin 'head' relative to actin.

Urea synthesis: ATP hydrolysis drives the ornithine cycle which removes toxic ammonia.

$2NH_3 + CO_2 + 3ATP + 3H_2O \rightarrow$ urea + AMP + 2ATP

Protein synthesis: ATP is used to 'load' amino acids onto transfer RNA.

Active transport systems are driven by the phosphorylation of membrane-bound proteins.

Calvin cycle (dark stage of photosynthesis): ATP hydrolysis drives the cyclic reduction of CO_2 to triose phosphate.

$3CO_2 + 6ATP \rightarrow$ triose phosphate

Nitrogen fixation involves the ATP-driven reduction of molecular nitrogen.

$N_2 + 8[H] + 12ATP \rightarrow 2NH_4^+ + 12 ADP + 12P$

Bioluminescence: ATP hydrolysis drives the oxidation of luciferin which releases some energy as visible light – useful for fireflies!

ATP: THE CENTRAL MOLECULE IN METABOLISM

Chemiosmosis: A proton gradient across an impermeable membrane is dissipated and the energy released is used to drive the phosphorylation of ADP.

H$^+$ (PROTONS)

ATP SYNTHETASE

MEMBRANE

ATP

ADP + P

H$^+$

PROTON PUMP: driven by energy from RESPIRATION OF FOODS (OXIDATIVE PHOSPHORYLATION) or from ABSORPTION OF LIGHT (PHOTOPHOSPHORYLATION)

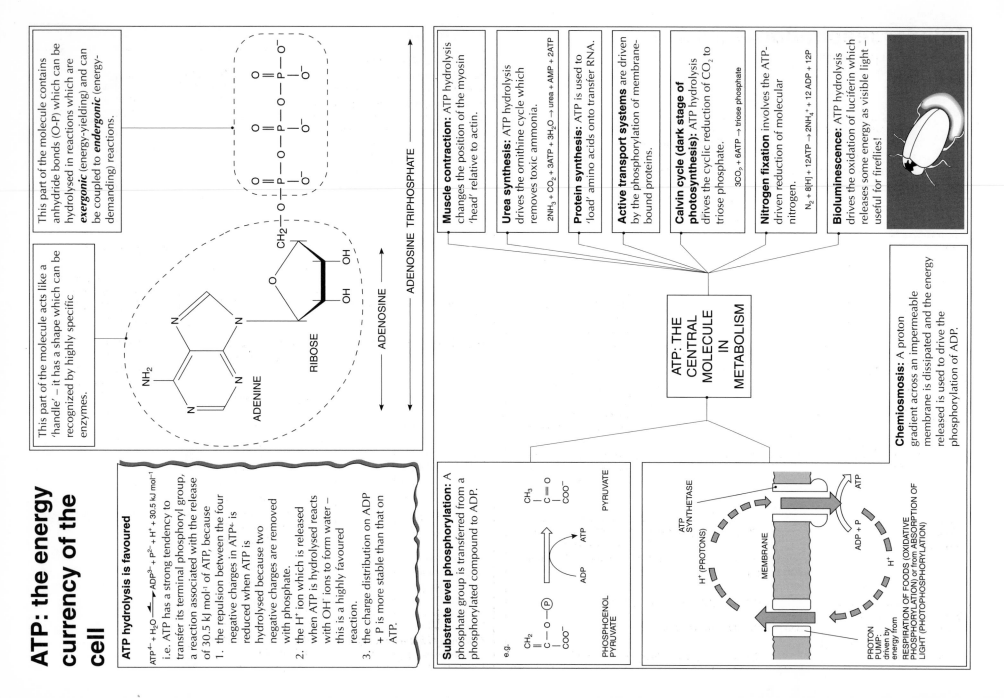

Chloroplasts: sites of photosynthesis contain photosensitive pigments

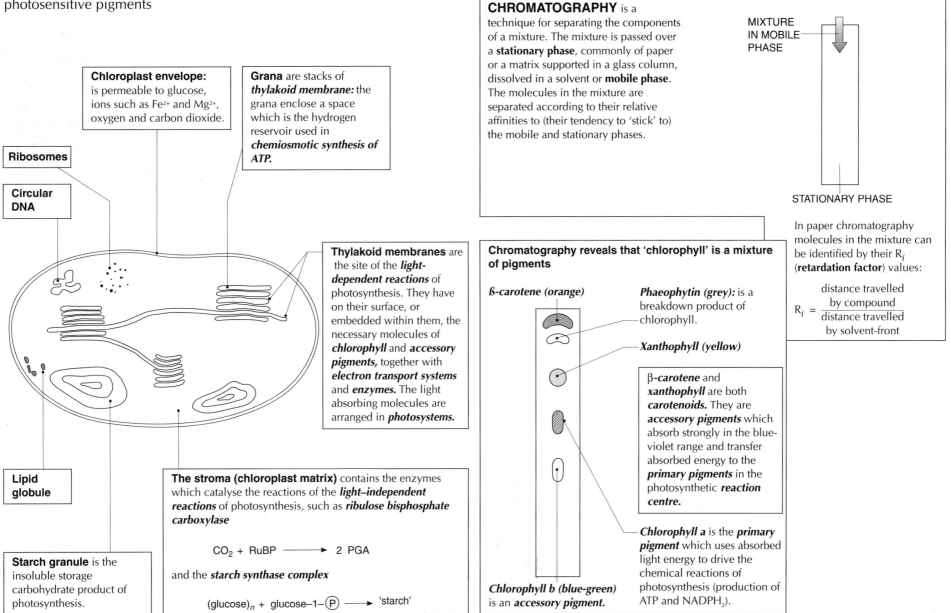

Chloroplast envelope: is permeable to glucose, ions such as Fe^{2+} and Mg^{2+}, oxygen and carbon dioxide.

Grana are stacks of **thylakoid membrane:** the grana enclose a space which is the hydrogen reservoir used in **chemiosmotic synthesis of ATP.**

Ribosomes

Circular DNA

Thylakoid membranes are the site of the **light-dependent reactions** of photosynthesis. They have on their surface, or embedded within them, the necessary molecules of **chlorophyll** and **accessory pigments,** together with **electron transport systems** and **enzymes.** The light absorbing molecules are arranged in **photosystems.**

Lipid globule

The stroma (chloroplast matrix) contains the enzymes which catalyse the reactions of the **light–independent reactions** of photosynthesis, such as **ribulose bisphosphate carboxylase**

$$CO_2 + RuBP \longrightarrow 2\ PGA$$

and the **starch synthase complex**

$$(glucose)_n + glucose\text{–}1\text{–}\circled{P} \longrightarrow \text{'starch'}$$

Starch granule is the insoluble storage carbohydrate product of photosynthesis.

CHROMATOGRAPHY is a technique for separating the components of a mixture. The mixture is passed over a **stationary phase**, commonly of paper or a matrix supported in a glass column, dissolved in a solvent or **mobile phase**. The molecules in the mixture are separated according to their relative affinities to (their tendency to 'stick' to) the mobile and stationary phases.

MIXTURE IN MOBILE PHASE

STATIONARY PHASE

In paper chromatography molecules in the mixture can be identified by their R_f (**retardation factor**) values:

$$R_f = \frac{\text{distance travelled by compound}}{\text{distance travelled by solvent-front}}$$

Chromatography reveals that 'chlorophyll' is a mixture of pigments

ß-carotene (orange)

Phaeophytin (grey): is a breakdown product of chlorophyll.

Xanthophyll (yellow)

β-**carotene** and **xanthophyll** are both **carotenoids.** They are **accessory pigments** which absorb strongly in the blue-violet range and transfer absorbed energy to the **primary pigments** in the photosynthetic **reaction centre.**

Chlorophyll a is the **primary pigment** which uses absorbed light energy to drive the chemical reactions of photosynthesis (production of ATP and $NADPH_2$).

Chlorophyll b (blue-green) is an **accessory pigment.**

Light reaction: non-cyclic photophosphorylation

Light energy excites electrons, resulting in the splitting of water and the synthesis of ATP and NADPH$_2$.

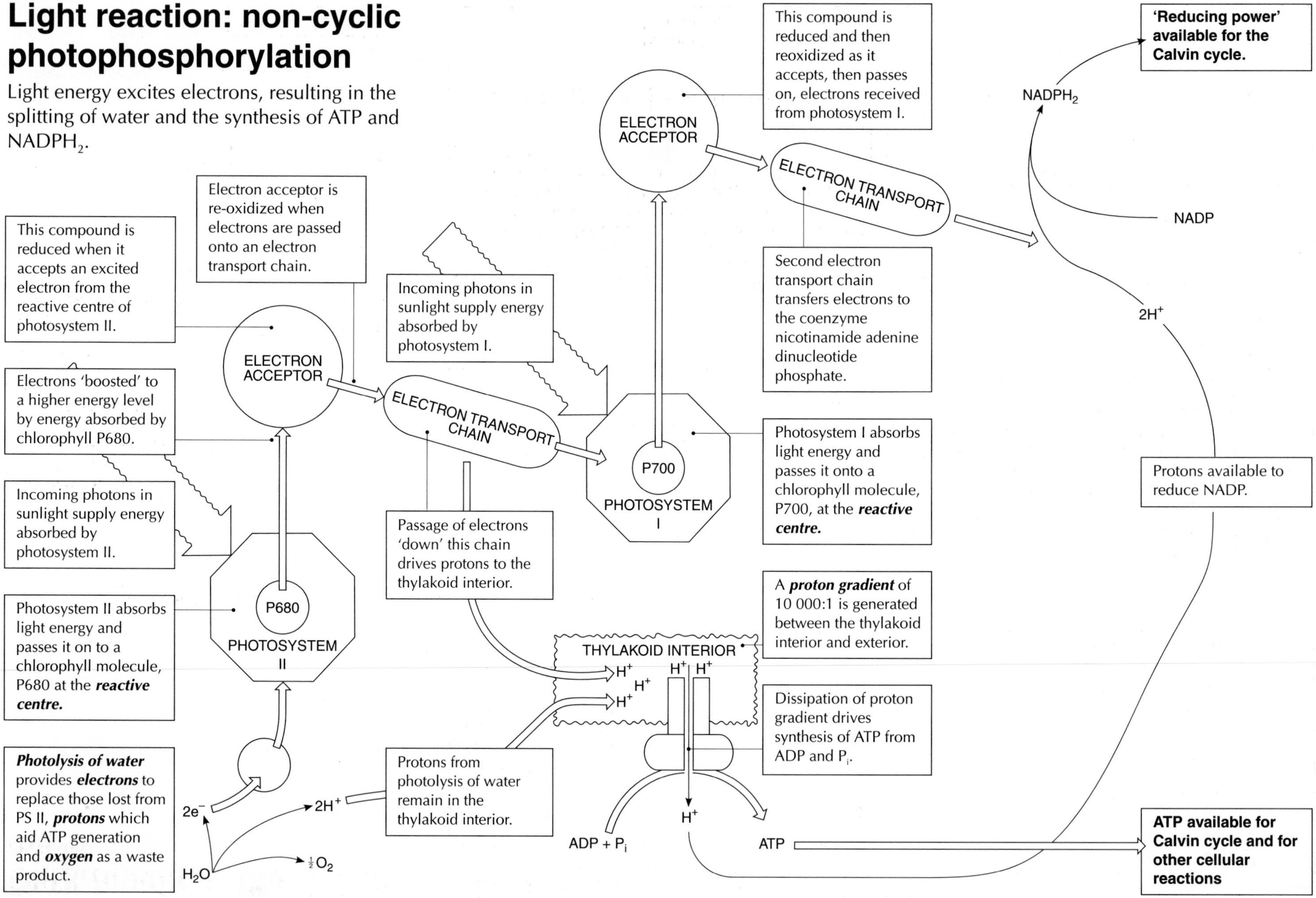

This compound is reduced and then reoxidized as it accepts, then passes on, electrons received from photosystem I.

'Reducing power' available for the Calvin cycle.

This compound is reduced when it accepts an excited electron from the reactive centre of photosystem II.

Electron acceptor is re-oxidized when electrons are passed onto an electron transport chain.

Incoming photons in sunlight supply energy absorbed by photosystem I.

ELECTRON ACCEPTOR

ELECTRON TRANSPORT CHAIN

NADPH$_2$

NADP

Second electron transport chain transfers electrons to the coenzyme nicotinamide adenine dinucleotide phosphate.

2H$^+$

Electrons 'boosted' to a higher energy level by energy absorbed by chlorophyll P680.

ELECTRON ACCEPTOR

ELECTRON TRANSPORT CHAIN

P700

PHOTOSYSTEM I

Photosystem I absorbs light energy and passes it onto a chlorophyll molecule, P700, at the **reactive centre.**

Incoming photons in sunlight supply energy absorbed by photosystem II.

Passage of electrons 'down' this chain drives protons to the thylakoid interior.

Protons available to reduce NADP.

Photosystem II absorbs light energy and passes it on to a chlorophyll molecule, P680 at the **reactive centre.**

P680

PHOTOSYSTEM II

A **proton gradient** of 10 000:1 is generated between the thylakoid interior and exterior.

THYLAKOID INTERIOR

H$^+$ H$^+$ H$^+$
H$^+$
H$^+$

Dissipation of proton gradient drives synthesis of ATP from ADP and P$_i$.

Photolysis of water provides **electrons** to replace those lost from PS II, **protons** which aid ATP generation and **oxygen** as a waste product.

2e$^-$

2H$^+$

Protons from photolysis of water remain in the thylakoid interior.

H$^+$

ADP + P$_i$

ATP

H$_2$O

$\frac{1}{2}$O$_2$

ATP available for Calvin cycle and for other cellular reactions

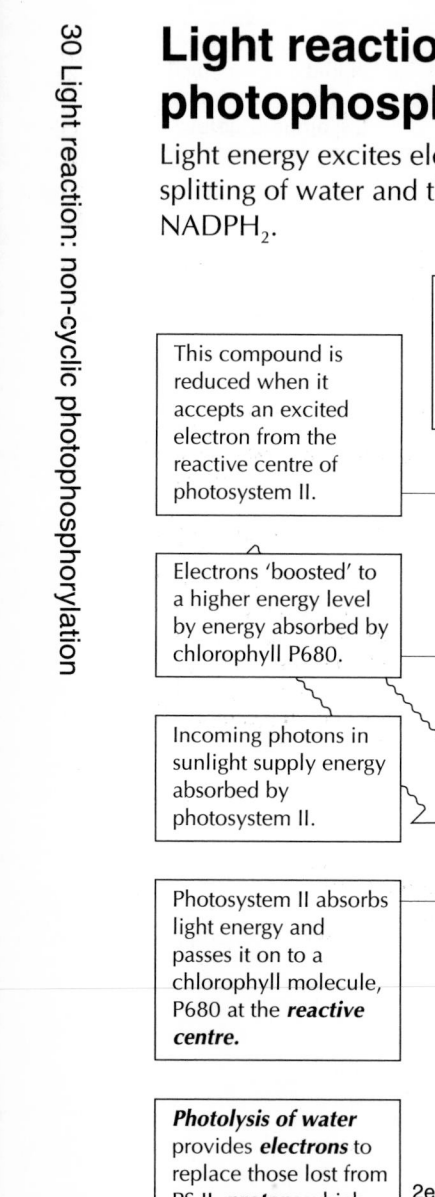

Dark reaction: the Calvin cycle

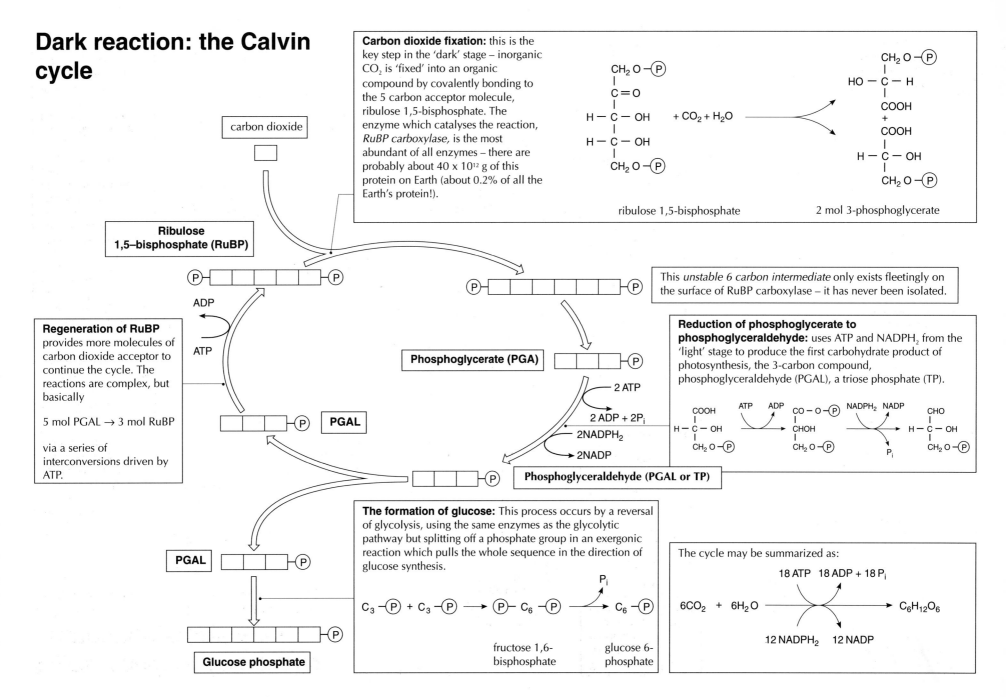

Carbon dioxide fixation: this is the key step in the 'dark' stage – inorganic CO_2 is 'fixed' into an organic compound by covalently bonding to the 5 carbon acceptor molecule, ribulose 1,5-bisphosphate. The enzyme which catalyses the reaction, *RuBP carboxylase,* is the most abundant of all enzymes – there are probably about 40×10^{12} g of this protein on Earth (about 0.2% of all the Earth's protein!).

ribulose 1,5-bisphosphate

2 mol 3-phosphoglycerate

carbon dioxide

Ribulose 1,5–bisphosphate (RuBP)

This *unstable 6 carbon intermediate* only exists fleetingly on the surface of RuBP carboxylase – it has never been isolated.

ADP

ATP

Regeneration of RuBP provides more molecules of carbon dioxide acceptor to continue the cycle. The reactions are complex, but basically

5 mol PGAL → 3 mol RuBP

via a series of interconversions driven by ATP.

Phosphoglycerate (PGA)

Reduction of phosphoglycerate to phosphoglyceraldehyde: uses ATP and $NADPH_2$ from the 'light' stage to produce the first carbohydrate product of photosynthesis, the 3-carbon compound, phosphoglyceraldehyde (PGAL), a triose phosphate (TP).

2 ATP

2 ADP + 2P_i

2NADPH$_2$

2NADP

PGAL

Phosphoglyceraldehyde (PGAL or TP)

PGAL

The formation of glucose: This process occurs by a reversal of glycolysis, using the same enzymes as the glycolytic pathway but splitting off a phosphate group in an exergonic reaction which pulls the whole sequence in the direction of glucose synthesis.

$C_3 - P + C_3 - P \longrightarrow P - C_6 - P \longrightarrow C_6 - P$

P_i

fructose 1,6-bisphosphate

glucose 6-phosphate

Glucose phosphate

The cycle may be summarized as:

18 ATP   18 ADP + 18 P_i

$6CO_2 + 6H_2O \longrightarrow C_6H_{12}O_6$

12 NADPH$_2$ 12 NADP

Nucleic acids I: DNA
Deoxyribonucleic acid

The Watson–Crick model for DNA suggests that the molecule is a double helix of two complementary, anti-parallel polynucleotide chains

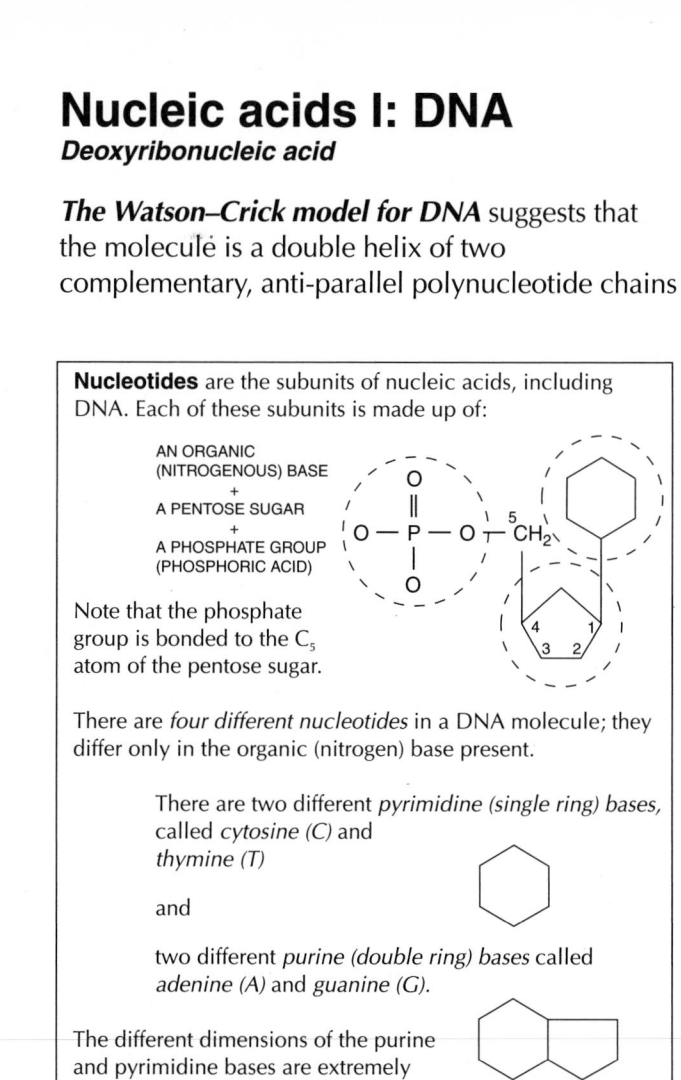

Nucleotides are the subunits of nucleic acids, including DNA. Each of these subunits is made up of:

AN ORGANIC
(NITROGENOUS) BASE
+
A PENTOSE SUGAR
+
A PHOSPHATE GROUP
(PHOSPHORIC ACID)

Note that the phosphate group is bonded to the C_5 atom of the pentose sugar.

There are *four different nucleotides* in a DNA molecule; they differ only in the organic (nitrogen) base present.

There are two different *pyrimidine (single ring) bases*, called *cytosine (C)* and *thymine (T)*

and

two different *purine (double ring) bases* called *adenine (A)* and *guanine (G)*.

The different dimensions of the purine and pyrimidine bases are extremely important in the formation of the double-stranded DNA molecule.

Nucleotides are linked to form a *polynucleotide* by the formation of *3' 5' phosphodiester links* in which a phosphate group forms a bridge between the C_3 of one sugar molecule and the C_5 of the next sugar molecule.

There are *ten base pairs* for each pitch of the double helix.

This is the 5' *end of the chain* since the terminal phosphate group is only bonded to the C_5 of the sugar molecule.

Base pairing in DNA was proposed to explain how two polynucleotide chains could be held together by hydrogen bonds. To accommodate the measured dimensions of the molecule each base pair comprises *one purine - one pyrimidine*.

The double helix is most stable, that is the greatest number of hydrogen bonds is formed, when the base pairs

A ⸬⸬⸬⸬ T   (two hydrogen bonds)

and   G ⸬⸬⸬⸬ C   (three hydrogen bonds)

are formed. These are *complementary base pairs*.

Note that in order to form and maintain this number of hydrogen bonds the nucleotides are inverted with respect to one another so that the phosphate groups (here shown as Ⓟ) *face in opposite directions*.

The chains are *anti-parallel,* that is one chain runs from 5'→3' whilst the other runs from 3'→5'.

The chains are *complementary:* because of base pairing, the base sequence on one of the chains automatically dictates the base sequence on the other.

This is the 3' *end of the chain* since the C_3 atom of the sugar molecule of the final nucleotide has a 'free' -OH group which is not part of a phosphodiester link.

Nucleic acids II: RNA

Ribonucleic acid has a number of functions in protein synthesis.

RNA is composed of nucleotides, but differs from DNA in that …

… and RNA is typically single-stranded, although it may assume complex structures depending on its function.

The pentose *ribose*, which has an -OH group at the 2' carbon atom, replaces *2'-deoxyribose*.

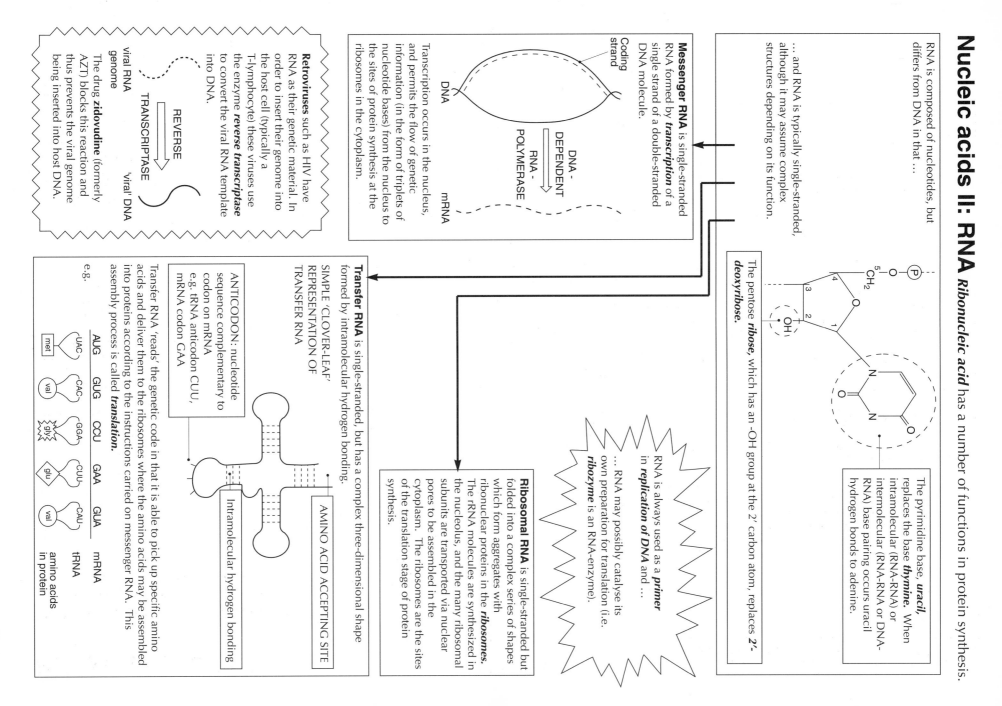

The pyrimidine base, *uracil*, replaces the base *thymine*. When intramolecular (RNA-RNA) or intermolecular (RNA-RNA or DNA-RNA) base pairing occurs uracil forms hydrogen bonds to adenine.

Messenger RNA is single-stranded RNA formed by *transcription* of a single strand of a double-stranded DNA molecule.

Coding strand

DNA

DNA-DEPENDENT RNA-POLYMERASE

mRNA

Transcription occurs in the nucleus, and permits the flow of genetic information (in the form of triplets of nucleotide bases) from the nucleus to the sites of protein synthesis at the ribosomes in the cytoplasm.

Retroviruses such as HIV have RNA as their genetic material. In order to insert their genome into the host cell (typically a T-lymphocyte) these viruses use the enzyme *reverse transcriptase* to convert the viral RNA template into DNA.

REVERSE TRANSCRIPTASE

viral RNA genome

'viral' DNA

The drug **zidovudine** (formerly AZT) blocks this reaction and thus prevents the viral genome being inserted into host DNA.

RNA is always used as a *primer* in *replication of DNA* and …

… RNA may possibly catalyse its own preparation for translation (i.e. *ribozyme* is an RNA-enzyme).

Ribosomal RNA is single-stranded but folded into a complex series of shapes which form aggregates with ribonuclear proteins in the *ribosomes*. The rRNA molecules are synthesized in the nucleolus, and the many ribosomal subunits are transported via nuclear pores to be assembled in the cytoplasm. The ribosomes are the sites of the translation stage of protein synthesis.

Transfer RNA is single-stranded, but has a complex three-dimensional shape formed by intramolecular hydrogen bonding.

SIMPLE 'CLOVER-LEAF' REPRESENTATION OF TRANSFER RNA

ANTICODON: nucleotide sequence complementary to codon on mRNA
e.g. tRNA anticodon CUU, mRNA codon GAA

Intramolecular hydrogen bonding

AMINO ACID ACCEPTING SITE

Transfer RNA 'reads' the genetic code in that it is able to pick up specific amino acids and deliver them to the ribosomes where the amino acids may be assembled into proteins according to the instructions carried on messenger RNA. This assembly process is called **translation.**

e.g.

AUG	GUG	CCU	GAA	GUA	mRNA
UAC	CAC	GGA	CUU	CAU	tRNA
met	val	gly	glu	val	amino acids in protein

DNA, genes and proteins

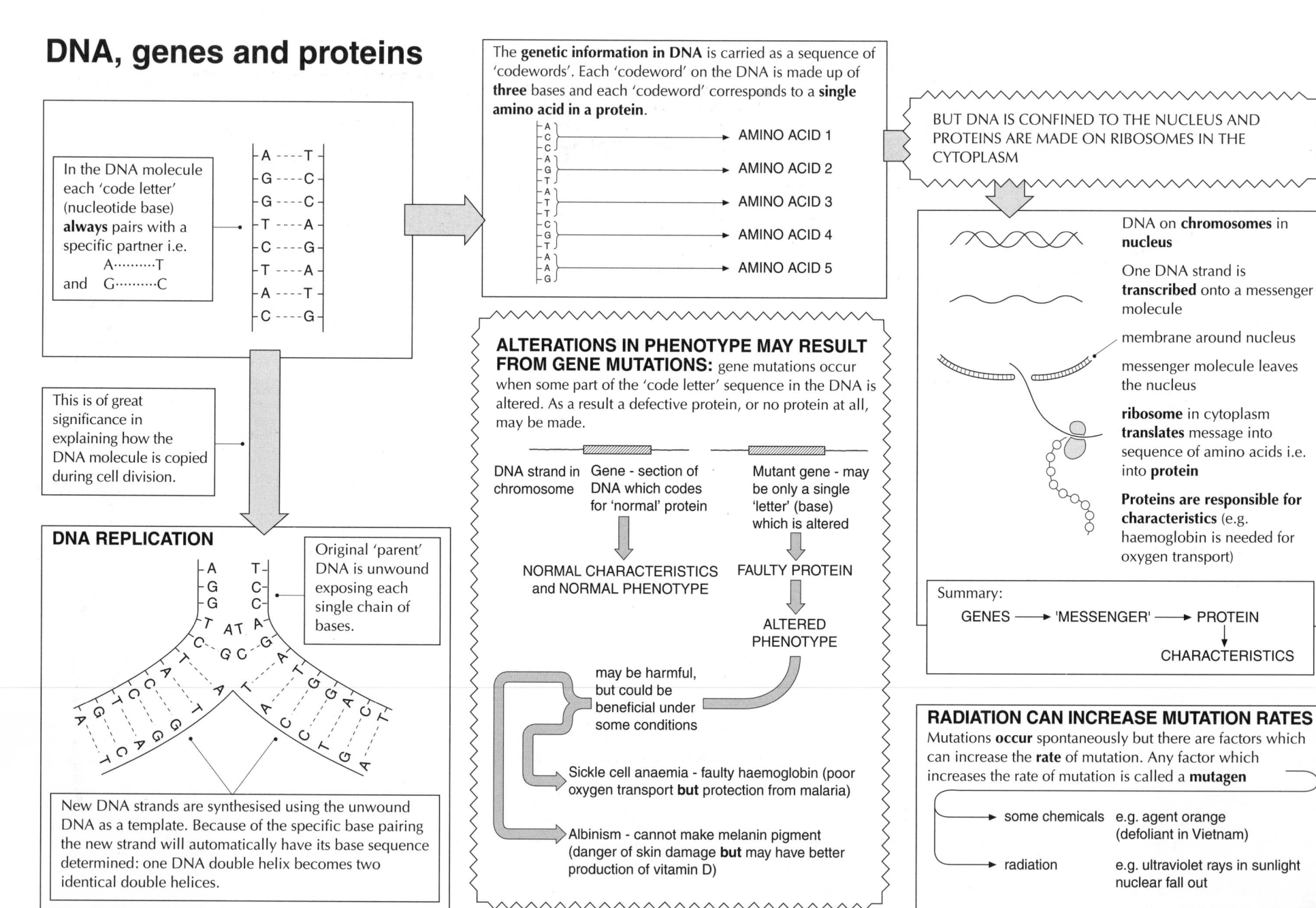

In the DNA molecule each 'code letter' (nucleotide base) **always** pairs with a specific partner i.e.

A·········T

and   G·········C

```
A ----T
G - - - C
G - - - C
T ----A
C ----G
T ----A
A ----T
C ----G
```

This is of great significance in explaining how the DNA molecule is copied during cell division.

DNA REPLICATION

Original 'parent' DNA is unwound exposing each single chain of bases.

New DNA strands are synthesised using the unwound DNA as a template. Because of the specific base pairing the new strand will automatically have its base sequence determined: one DNA double helix becomes two identical double helices.

The **genetic information in DNA** is carried as a sequence of 'codewords'. Each 'codeword' on the DNA is made up of **three** bases and each 'codeword' corresponds to a **single amino acid in a protein**.

- AMINO ACID 1
- AMINO ACID 2
- AMINO ACID 3
- AMINO ACID 4
- AMINO ACID 5

ALTERATIONS IN PHENOTYPE MAY RESULT FROM GENE MUTATIONS:
gene mutations occur when some part of the 'code letter' sequence in the DNA is altered. As a result a defective protein, or no protein at all, may be made.

DNA strand in chromosome

Gene - section of DNA which codes for 'normal' protein

Mutant gene - may be only a single 'letter' (base) which is altered

NORMAL CHARACTERISTICS and NORMAL PHENOTYPE

FAULTY PROTEIN

ALTERED PHENOTYPE

may be harmful, but could be beneficial under some conditions

Sickle cell anaemia - faulty haemoglobin (poor oxygen transport **but** protection from malaria)

Albinism - cannot make melanin pigment (danger of skin damage **but** may have better production of vitamin D)

BUT DNA IS CONFINED TO THE NUCLEUS AND PROTEINS ARE MADE ON RIBOSOMES IN THE CYTOPLASM

DNA on **chromosomes** in **nucleus**

One DNA strand is **transcribed** onto a messenger molecule

membrane around nucleus

messenger molecule leaves the nucleus

ribosome in cytoplasm **translates** message into sequence of amino acids i.e. into **protein**

Proteins are responsible for characteristics (e.g. haemoglobin is needed for oxygen transport)

Summary:

GENES ⟶ 'MESSENGER' ⟶ PROTEIN

↓

CHARACTERISTICS

RADIATION CAN INCREASE MUTATION RATES

Mutations **occur** spontaneously but there are factors which can increase the **rate** of mutation. Any factor which increases the rate of mutation is called a **mutagen**

→ some chemicals   e.g. agent orange (defoliant in Vietnam)

→ radiation   e.g. ultraviolet rays in sunlight nuclear fall out

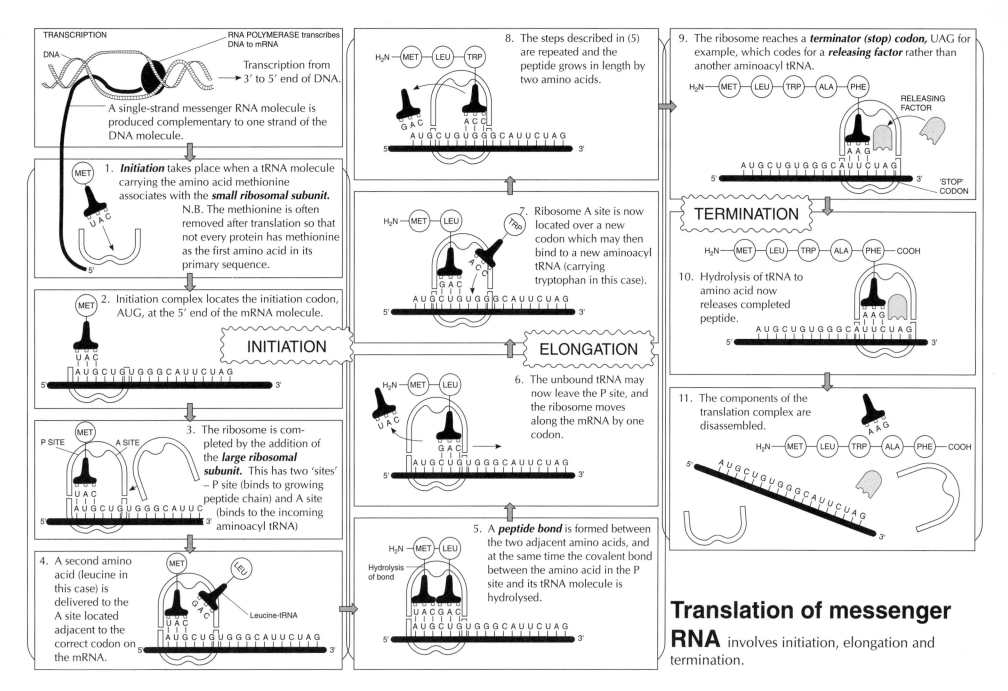

TRANSCRIPTION

RNA POLYMERASE transcribes DNA to mRNA

DNA

Transcription from 3' to 5' end of DNA.

A single-strand messenger RNA molecule is produced complementary to one strand of the DNA molecule.

1. **Initiation** takes place when a tRNA molecule carrying the amino acid methionine associates with the **small ribosomal subunit.** N.B. The methionine is often removed after translation so that not every protein has methionine as the first amino acid in its primary sequence.

MET
U A C
5'

2. Initiation complex locates the initiation codon, AUG, at the 5' end of the mRNA molecule.

MET
U A C
U A C
AUGCUGUGGGCAUUCUAG
5'   3'

INITIATION

3. The ribosome is completed by the addition of the **large ribosomal subunit.** This has two 'sites' – P site (binds to growing peptide chain) and A site (binds to the incoming aminoacyl tRNA)

P SITE   MET   A SITE
U A C
U A C
AUGCUGUGGGCAUUC
5'   3'

4. A second amino acid (leucine in this case) is delivered to the A site located adjacent to the correct codon on the mRNA.

MET   LEU
G A C
U A C
AUGCUGUGGGCAUUCUAG
5'   3'
Leucine-tRNA

8. The steps described in (5) are repeated and the peptide grows in length by two amino acids.

H₂N—MET—LEU—TRP
G A C   A C C
AUGCUGUGGGCAUUCUAG
5'   3'

7. Ribosome A site is now located over a new codon which may then bind to a new aminoacyl tRNA (carrying tryptophan in this case).

H₂N—MET—LEU   TRP
A C C
G A C
AUGCUGUGGGCAUUCUAG
5'   3'

ELONGATION

6. The unbound tRNA may now leave the P site, and the ribosome moves along the mRNA by one codon.

H₂N—MET—LEU
U A C
G A C
AUGCUGUGGGCAUUCUAG
5'   3'

5. A **peptide bond** is formed between the two adjacent amino acids, and at the same time the covalent bond between the amino acid in the P site and its tRNA molecule is hydrolysed.

H₂N—MET—LEU
Hydrolysis of bond
U A C G A C
AUGCUGUGGGCAUUCUAG
5'   3'

9. The ribosome reaches a **terminator (stop) codon,** UAG for example, which codes for a **releasing factor** rather than another aminoacyl tRNA.

H₂N—MET—LEU—TRP—ALA—PHE
RELEASING FACTOR
A A G
AUGCUGUGGGCAUUCUAG
5'   3'   'STOP' CODON

TERMINATION

H₂N—MET—LEU—TRP—ALA—PHE—COOH

10. Hydrolysis of tRNA to amino acid now releases completed peptide.

A A G
AUGCUGUGGGCAUUCUAG
5'   3'

11. The components of the translation complex are disassembled.

A A G
H₂N—MET—LEU—TRP—ALA—PHE—COOH

5'
AUGCUGUGGGCAUUCUAG
3'

Translation of messenger RNA involves initiation, elongation and termination.

Genes control cell characteristics

DNA $\xrightarrow{\text{TRANSCRIPTION}}$ RNA $\xrightarrow{\text{TRANSLATION}}$ PROTEIN $\longrightarrow$ CHARACTERISTICS

DNA in nucleus contains the *genes* (specific sequences of bases which code for particular proteins), *regulators* (which switch genes 'on' or 'off') and many stretches with no apparent function (*junk* or *redundant DNA*).

Transcription is the synthesis of an RNA molecule with a base sequence complementary to a section of DNA. The process is catalysed by a *DNA dependent RNA polymerase* which first binds to a promoter region on the DNA and then adds nucleotide triphosphates from the nuclear cytoplasm to form an mRNA chain which lengthens in the 5' → 3' direction.

Transcription of a gene produces a *primary mRNA transcript* which undergoes further processing (including the removal of some redundant base sequences – introns – and the addition of some signalling sequences – caps) to produce *mature messenger RNA*.

N.B. RNA contains the base uracil (U) in place of thymine (T) so that mRNA has U opposite any A in the complementary DNA.

Three different RNA products are generated by transcription – *transfer RNA*, *ribosomal RNA* and *messenger RNA*.

Nucleus

Nuclear envelope with protein-lined pores

Cytoplasm

Structural proteins

Enzymes regulate synthesis of

Carbohydrates

Lipids

Nucleic acids

Other organic molecules

Translation of the mature mRNA matches *codons* (triplets of nucleotide bases) on mRNA to *anticodons* on tRNA, which means that amino acids are assembled into *polypeptides* at the intact ribosome.

Transfer RNA collects amino acids from the 'pool' of amino acids dissolved in the cytoplasm, which originate from food, or transamination in the liver.

Ribosomal RNA combines with ribosomal proteins to form small and large ribosomal subunits. In the presence of the cytoplasmic *initiation factors* and a mature RNA the subunits assemble into a *ribosome*.

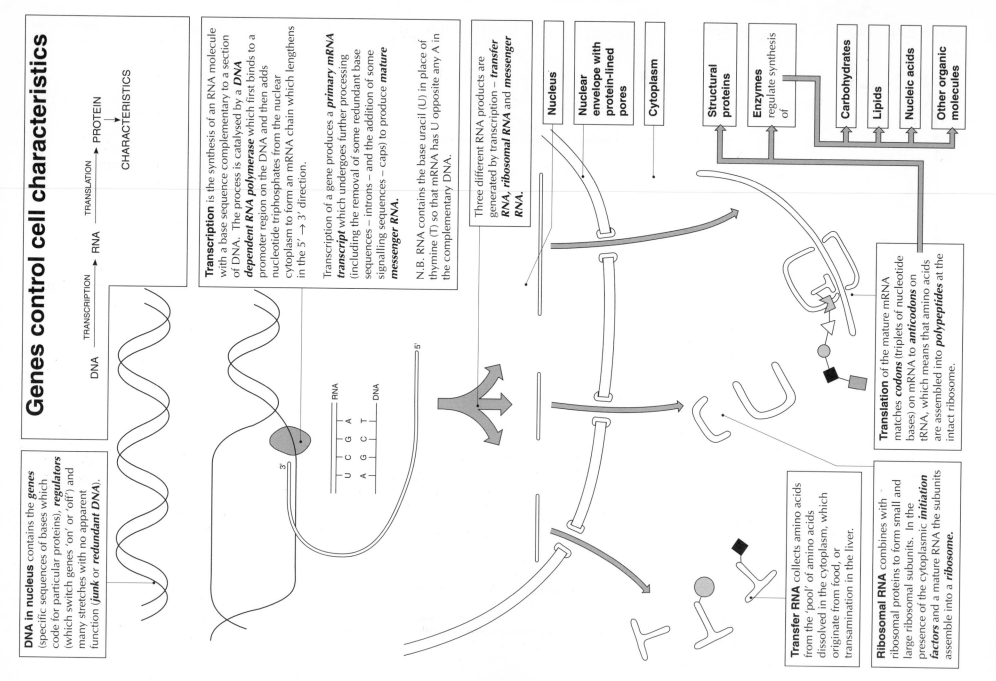

36 Genes control cell characteristics

Genetic engineering (recombinant DNA technology)

depends on enzymes and culture of micro-organisms

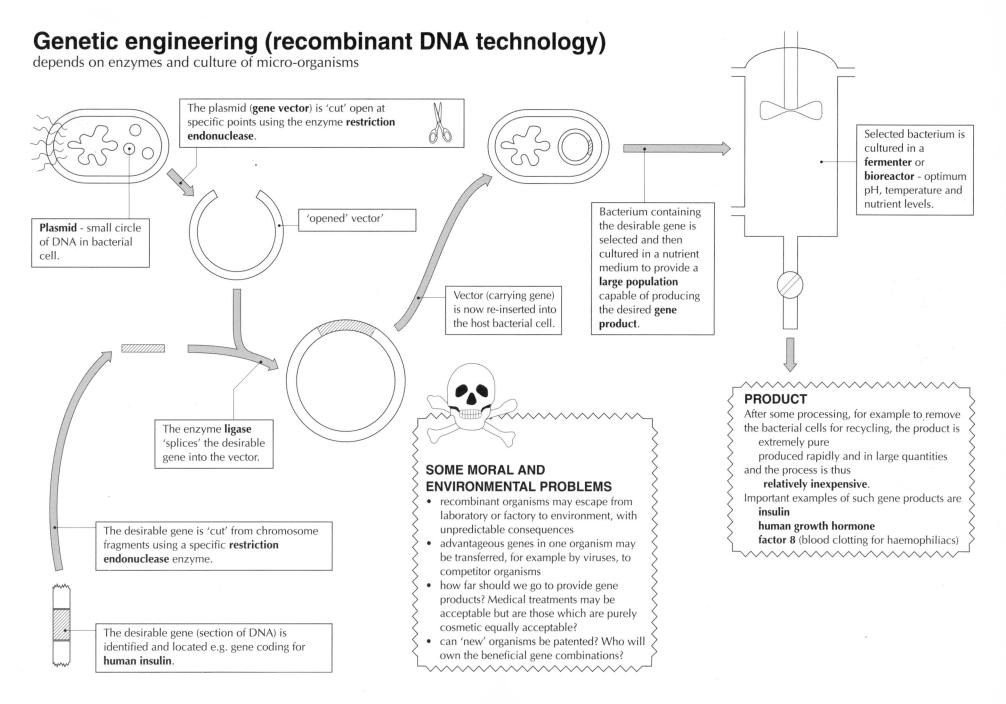

The plasmid (**gene vector**) is 'cut' open at specific points using the enzyme **restriction endonuclease**.

Plasmid - small circle of DNA in bacterial cell.

'opened' vector'

The enzyme **ligase** 'splices' the desirable gene into the vector.

The desirable gene is 'cut' from chromosome fragments using a specific **restriction endonuclease** enzyme.

The desirable gene (section of DNA) is identified and located e.g. gene coding for **human insulin**.

Vector (carrying gene) is now re-inserted into the host bacterial cell.

Bacterium containing the desirable gene is selected and then cultured in a nutrient medium to provide a **large population** capable of producing the desired **gene product**.

Selected bacterium is cultured in a **fermenter** or **bioreactor** - optimum pH, temperature and nutrient levels.

PRODUCT

After some processing, for example to remove the bacterial cells for recycling, the product is extremely pure produced rapidly and in large quantities and the process is thus **relatively inexpensive**.

Important examples of such gene products are

insulin

human growth hormone

factor 8 (blood clotting for haemophiliacs)

SOME MORAL AND ENVIRONMENTAL PROBLEMS

- recombinant organisms may escape from laboratory or factory to environment, with unpredictable consequences
- advantageous genes in one organism may be transferred, for example by viruses, to competitor organisms
- how far should we go to provide gene products? Medical treatments may be acceptable but are those which are purely cosmetic equally acceptable?
- can 'new' organisms be patented? Who will own the beneficial gene combinations?

Genetic profiling: minisatellites and probes

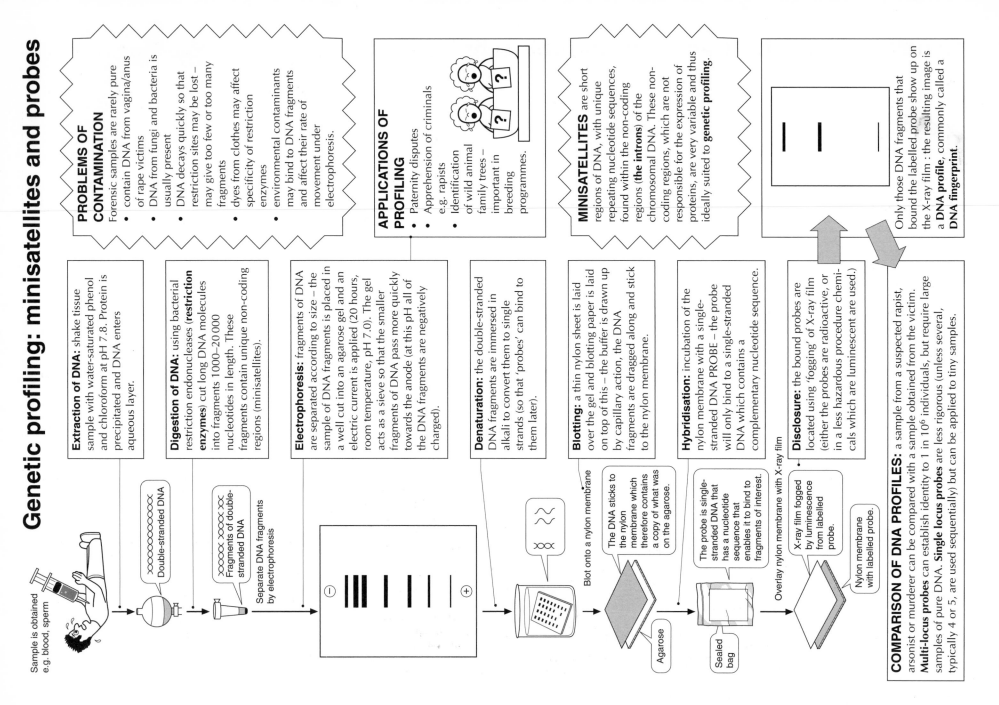

PROBLEMS OF CONTAMINATION

Forensic samples are rarely pure

- contain DNA from vagina/anus of rape victims
- DNA from fungi and bacteria is usually present
- DNA decays quickly so that restriction sites may be lost – may give too few or too many fragments
- dyes from clothes may affect specificity of restriction enzymes
- environmental contaminants may bind to DNA fragments and affect their rate of movement under electrophoresis.

APPLICATIONS OF PROFILING

- Paternity disputes
- Apprehension of criminals e.g. rapists
- Identification of wild animal family trees – important in breeding programmes.

MINISATELLITES are short regions of DNA, with unique repeating nucleotide sequences, found within the non-coding regions **(the introns)** of the chromosomal DNA. These non-coding regions, which are not responsible for the expression of proteins, are very variable and thus ideally suited to **genetic profiling.**

Only those DNA fragments that bound the labelled probe show up on the X-ray film : the resulting image is a **DNA profile,** commonly called a **DNA fingerprint.**

Extraction of DNA: shake tissue sample with water-saturated phenol and chloroform at pH 7.8. Protein is precipitated and DNA enters aqueous layer.

Digestion of DNA: using bacterial restriction endonucleases **(restriction enzymes)** cut long DNA molecules into fragments 1000–20 000 nucleotides in length. These fragments contain unique non-coding regions (minisatellites).

Electrophoresis: fragments of DNA are separated according to size – the sample of DNA fragments is placed in a well cut into an agarose gel and an electric current is applied (20 hours, room temperature, pH 7.0). The gel acts as a sieve so that the smaller fragments of DNA pass more quickly towards the anode (at this pH all of the DNA fragments are negatively charged).

Denaturation: the double-stranded DNA fragments are immersed in alkali to convert them to single strands (so that 'probes' can bind to them later).

Blotting: a thin nylon sheet is laid over the gel and blotting paper is laid on top of this – the buffer is drawn up by capillary action, the DNA fragments are dragged along and stick to the nylon membrane.

Hybridisation: incubation of the nylon membrane with a single-stranded DNA PROBE – the probe will only bind to a single-stranded DNA which contains a complementary nucleotide sequence.

Disclosure: the bound probes are located using 'fogging' of X-ray film (either the probes are radioactive, or in a less hazardous procedure chemicals which are luminescent are used.)

Sample is obtained e.g. blood, sperm

Double-stranded DNA

Fragments of double-stranded DNA

Separate DNA fragments by electrophoresis

Blot onto a nylon membrane

The DNA sticks to the nylon membrane which therefore contains a copy of what was on the agarose.

Agarose

The probe is single-stranded DNA that has a nucleotide sequence that enables it to bind to fragments of interest.

Sealed bag

Overlay nylon membrane with X-ray film

X-ray film fogged by luminescence from labelled probe.

Nylon membrane with labelled probe.

COMPARISON OF DNA PROFILES: a sample from a suspected rapist, arsonist or murderer can be compared with a sample obtained from the victim. **Multi-locus probes** can establish identity to 1 in 10^6 individuals, but require large samples of pure DNA. **Single locus probes** are less rigorous (unless several, typically 4 or 5, are used sequentially) but can be applied to tiny samples.

Cystic fibrosis is caused by faulty ion transport, and is the most common inherited fatal disease in Caucasian populations.

Screening for CF involves identification of the defective gene. The CFTR gene is located on human chromosome 7, but direct screening for mutant forms is complicated since there are many mutations of this gene, although it is possible for the most common (70%) of the mutation. This is called ΔF508 since it results in the omission of one phenylalanine molecule from position 508 in the protein chain. The direct screening is usually carried out on cells from a buccal smear, on white blood cells, or on chorionic villus/amniotic fluid cells if the test is pre-natal. **Indirect screening** uses gene probes to detect DNA markers close to less common CFTR mutations, but with more than 100 known mutations infallible testing is financially prohibitive.

Gene therapy offers considerable hope as a cure for CF. This therapy would introduce healthy CFTR genes into defective epithelial cells which could then synthesise functional protein resulting in a normal flow of chloride ions and water across these membranes. Success depends on identification of the faulty CFTR gene, large scale synthesis of replacement 'normal' CFTR gene, and development of vector systems to transfer the healthy gene into the DNA of target epithelial cells.

Gametogenesis is affected – CF adults are usually sterile or of very low fertility.

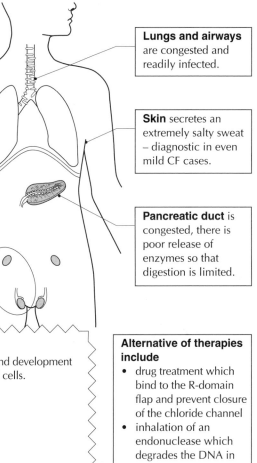

Lungs and airways are congested and readily infected.

Skin secretes an extremely salty sweat – diagnostic in even mild CF cases.

Pancreatic duct is congested, there is poor release of enzymes so that digestion is limited.

The basic principle is

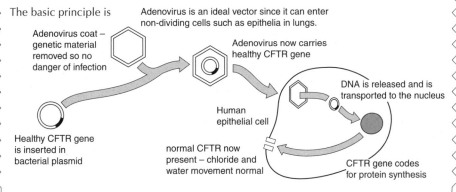

Adenovirus is an ideal vector since it can enter non-dividing cells such as epithelia in lungs.

Adenovirus coat – genetic material removed so no danger of infection

Adenovirus now carries healthy CFTR gene

Human epithelial cell

DNA is released and is transported to the nucleus

Healthy CFTR gene is inserted in bacterial plasmid

normal CFTR now present – chloride and water movement normal

CFTR gene codes for protein synthesis

Alternative of therapies include

- drug treatment which bind to the R-domain flap and prevent closure of the chloride channel
- inhalation of an endonuclease which degrades the DNA in the mucus and reduces its viscosity
- infusion of airways with calcium ions which appear to make alternative ion channels accessible to chloride ions

Patients suffering from CF have an ionic imbalance across the membranes of epithelial cells. This imbalance is caused by the absence of the normal form of a protein – **Cystic Fibrosis Transmembrane Regulator (CFTR)** – which controls the passage of chloride ions across these cell membranes.

Normal cell: chloride ions flow out, are followed by water, and lining of passages are kept moist with a free-moving layer of mucus.

CF cell: chloride movement is restricted, water does not enter the passages which become clogged with a drier, more viscous mucus.

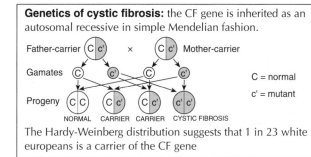

The presence of this mucus restricts the movement of materials along body passages, for example, breathing is difficult as airflow is restricted through the respiratory tree.

The situation is made worse because long fibres of DNA are released from the epithilial cells and from white blood cells in the locality. These fibres make the mucus even more viscous and difficult to move – physiotherapy involving lung percussion is only partially successful.

The thick, stagnant mucus offers a nutrient source for bacteria. These invade rapidly and cause severe lung infections – most CF sufferers have died of lung infections by their mid-twenties. One common bacterial coloniser is *Pseudomonas aeruginosa* – this bacterium has a thick capsule which further restricts the flow of mucus. CF sufferers try to restrict infection with a wide antibiotic regime.

Genetics of cystic fibrosis: the CF gene is inherited as an autosomal recessive in simple Mendelian fashion.

Father-carrier (C c') × (C c') Mother-carrier

Gamates (C) (c') (C) (c')

Progeny (C C) (C c') (C c') (c' c')
NORMAL   CARRIER CARRIER CYSTIC FIBROSIS

C = normal
c' = mutant

The Hardy-Weinberg distribution suggests that 1 in 23 white europeans is a carrier of the CF gene

The nucleus is the control centre of the cell

NUCLEOLUS is the site of manufacture of ribosomal subunits. Within the nucleolus are **nucleolar organisers** which contain multiple copies of the genes which are transcribed to ribosomal RNA.

The nucleolus disperses in preparation for nuclear division, and is reassembled at the end of telophase.

NUCLEOPLASM contains a variety of solutes, including ATP (energy source) and other nucleoside triphosphates which are the raw materials for DNA replication. The enzyme complex which regulates the replication of DNA (DNA polymerase) is also found here, as are ribosomal proteins awaiting assembly with ribosomal RNA into ribosomal subunits.

NUCLEAR PORE can regulate the entry (e.g. ribosomal proteins, nucleotides) and exit (e.g., ribosomal subunits, messenger RNA) of molecules to and from the nucleus. There is a precise and highly organised arrangement of fibrous proteins forming the nuclear pore complex to carry out this controlled transport.

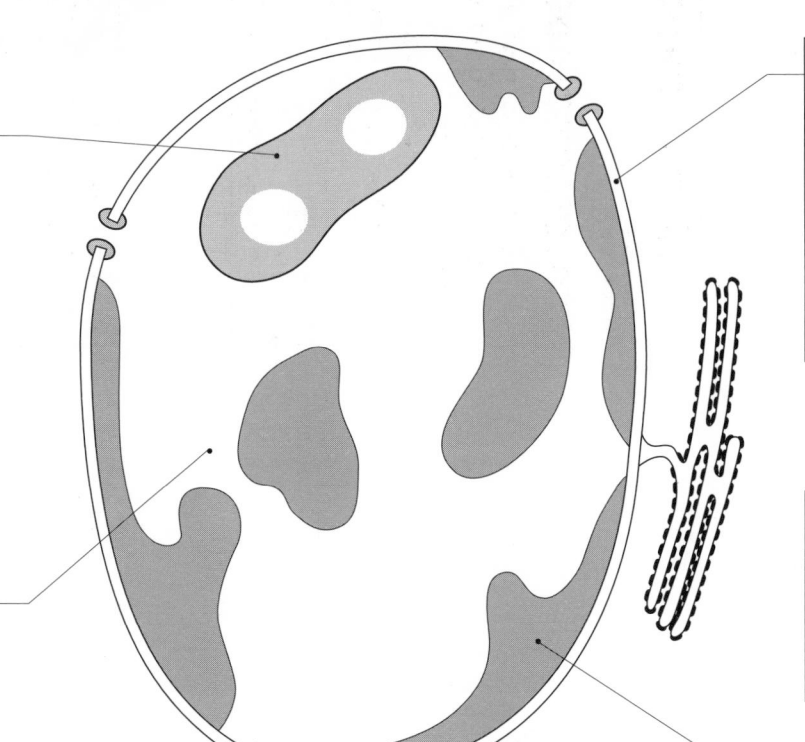

NUCLEAR ENVELOPE is a double membrane, the inner of which is lined by a protein mesh (the **lamina**) which may help to maintain the structure of the nucleus and to anchor the chromosomes, and the outer of which is continuous with the endoplasmic reticulum. The nuclear envelope is pierced by a number of **nuclear pores**.

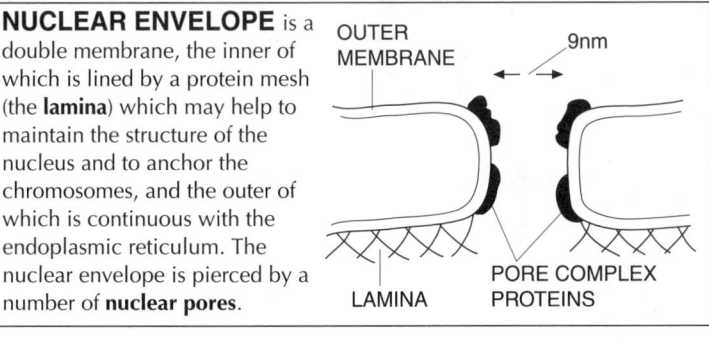

OUTER MEMBRANE

9nm

PORE COMPLEX PROTEINS

LAMINA

The presence of a nucleus is typical of **eukaryotic** cells. The divergence of the more primitive **prokaryotes** (no nuclear membrane) and the **eukaryotes** (the nuclear membrane present) in evolutionary history approximately coincides with the development of resistance to endosymbiosis. It may be that the nuclear membrane has been developed to protect the genetic material from the cytoplasmic enzymes which might be mobilised to digest invading organic matter.

CHROMATIN is the genetic material containing the coded information for protein synthesis in the cell. It is composed of DNA bound to basic proteins called **histones**. The DNA and histone are organised into **nucleosomes**.

During nuclear division the chromatin condenses to form the **chromosomes**, and the chromatin containing DNA which is being 'expressed' (transcribed into mRNA) becomes visible as more loosely coiled threads called **euchromatin**.

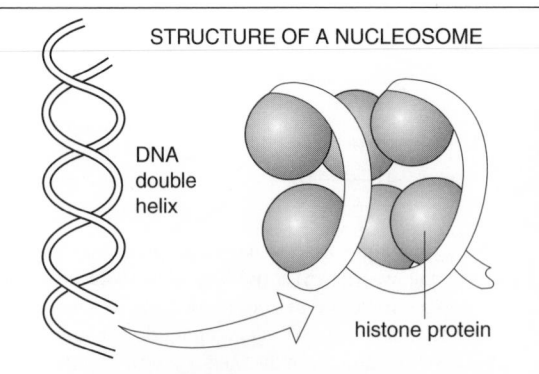

STRUCTURE OF A NUCLEOSOME

DNA double helix

histone protein

DNA replication and chromosomes

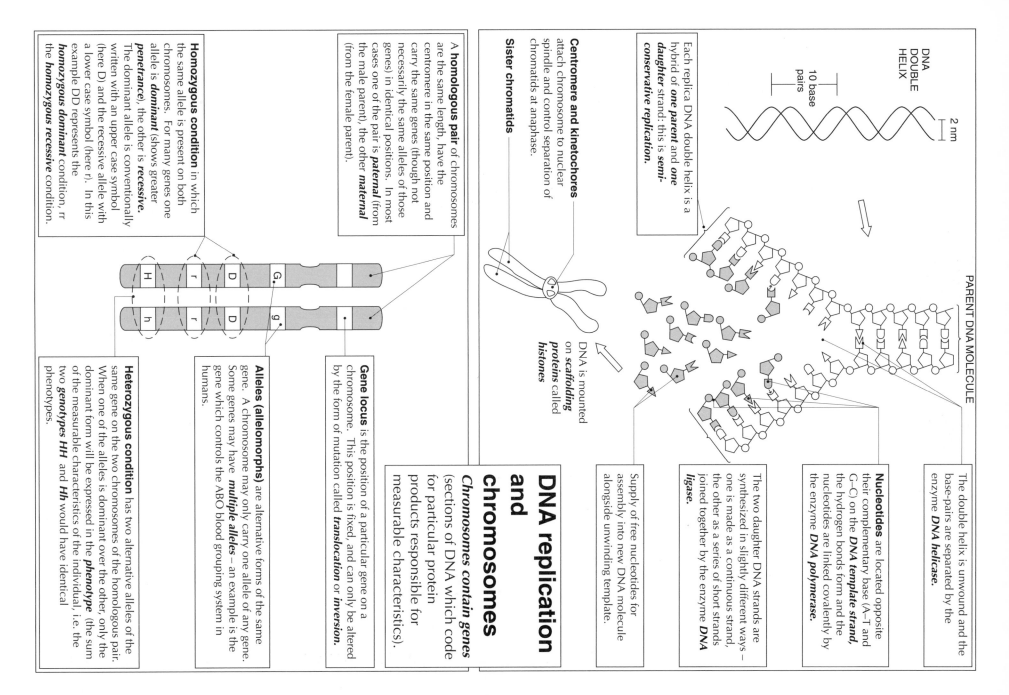

PARENT DNA MOLECULE

DNA DOUBLE HELIX

10 base pairs

2 nm

Each replica DNA double helix is a hybrid of **one parent** and **one daughter** strand: this is **semi-conservative replication.**

The double helix is unwound and the base-pairs are separated by the enzyme **DNA helicase.**

Nucleotides are located opposite their complementary base (A–T and C–G) on the **DNA template strand,** the hydrogen bonds form and the nucleotides are linked covalently by the enzyme **DNA polymerase.**

The two daughter DNA strands are synthesized in slightly different ways – one is made as a continuous strand, the other as a series of short strands joined together by the enzyme **DNA ligase.**

Supply of free nucleotides for assembly into new DNA molecule alongside unwinding template.

DNA is mounted on **scaffolding proteins** called **histones**

Centromere and kinetochores
attach chromosome to nuclear spindle and control separation of chromatids at anaphase.

Sister chromatids

DNA replication and chromosomes

Chromosomes contain genes (sections of DNA which code for particular protein products responsible for measurable characteristics).

A **homologous pair** of chromosomes are the same length, have the centromere in the same position and carry the same genes (though not necessarily the same alleles of those genes) in identical positions. In most cases one of the pair is **paternal** (from the male parent), the other **maternal** (from the female parent).

Gene locus is the position of a particular gene on a chromosome. This position is fixed, and can only be altered by the form of mutation called **translocation** or **inversion.**

Alleles (allelomorphs) are alternative forms of the same gene. A chromosome may only carry one allele of any gene. Some genes may have **multiple alleles** – an example is the gene which controls the ABO blood grouping system in humans.

Heterozygous condition has two alternative alleles of the same gene on the two chromosomes of the homologous pair. When one of the alleles is dominant over the other, only the dominant form will be expressed in the **phenotype** (the sum of the measurable characteristics of the individual, i.e. the two **genotypes HH** and **Hh** would have identical phenotypes.

Homozygous condition in which the same allele is present on both chromosomes. For many genes one allele is **dominant** (shows greater **penetrance**), the other is **recessive.** The dominant allele is conventionally written with an upper case symbol (here D) and the recessive allele with a lower case symbol (here r). In this example DD represents the **homozygous dominant** condition, rr the **homozygous recessive** condition.

H | r | D
h | r | D
G
g

Mitosis and growth

The significance of mitosis is that it involves duplication of the genetic material and its equal distribution to each of two 'daughter' cells: **variation is minimal.**

Importance of mitosis

1. It is the process which provides the cells required for the **growth** of zygote into a functioning multicellular organism – this requires an increase in number of cells from one to 6×10^{13} in humans.
2. It supplies the cells to **repair** worn-out or damaged tissues. In the human the replacement of skin, gut and lung linings and blood cells requires about 1×10^{11} cells per day.
3. It maintains the chromosome number. Daughter cells have identical sets of chromosomes and so function harmoniously as part of the tissue, organ or organism.
4. **Asexual reproduction** provides offspring which are genetically identical to the parent – ideal when rapidly establishing a population. Mitosis provides the cells which make up the fragments of the parent body dispersed during this form of reproduction.

The cell cycle
typically lasts from 8 to 24 h in humans – the nuclear division (mitosis) occupies about 10% of this time.

Interphase
DNA only visible as indistinct mass of **chromatin.** Nucleolus and nuclear membrane still intact. Centrioles lie close to one another.

As **interphase** moves to **prophase**

DNA undergoes **spiralization** and **replication.** Each chromosome is now **two identical chromatids,** held together at the centromere. Nucleolus and nuclear membrane disintegrate. **Centrioles** move to opposite poles, forming a **spindle** of **microtubules.**

The rate of replication may be extremely high – in humans 1×10^{11} m of DNA are produced per day (this is almost 700 miles per second!).

At **metaphase**

Chromosomes now attached to spindle at **kinetochore** on the centromere. The chromosomes are arranged in such a way that one chromatid from each pair lies on each side of the **equator.**

VINCRISTINE and VINBLASTINE are anti-cancer drugs which inhibit the formation of the spindle of microtubules.

Anaphase
precedes **telophase**

Centromere divides and spindle fibres contract to pull **chromatids** to opposite poles. The early separation of the chromatids constitutes **anaphase,** and the separation is complete (so that the chromatids are now **chromosomes**) when the spindle disintegrates and the nuclear membrane reforms at **telophase.**

During **cytokinesis** the tetraploid ($4n$) cell is 'pinched' into two 'daughter cells'. Each product has a **DNA content equal to the other and to the parent cell.** In animal cells the separation is brought about by two contractile proteins which form a **cleavage furrow;** in plant cells a **cell plate** is laid down and covered with cellulose to form a separating **cell wall.**

Important sites of mitosis in humans include gut epithelium and bone marrow – these rapidly-dividing cells are susceptible to non-specific anti-cancer drugs and thus patients undergoing treatment with these cytotoxic compounds often experience side effects which include mouth ulceration, irritation of the gut and reduced blood cell production.

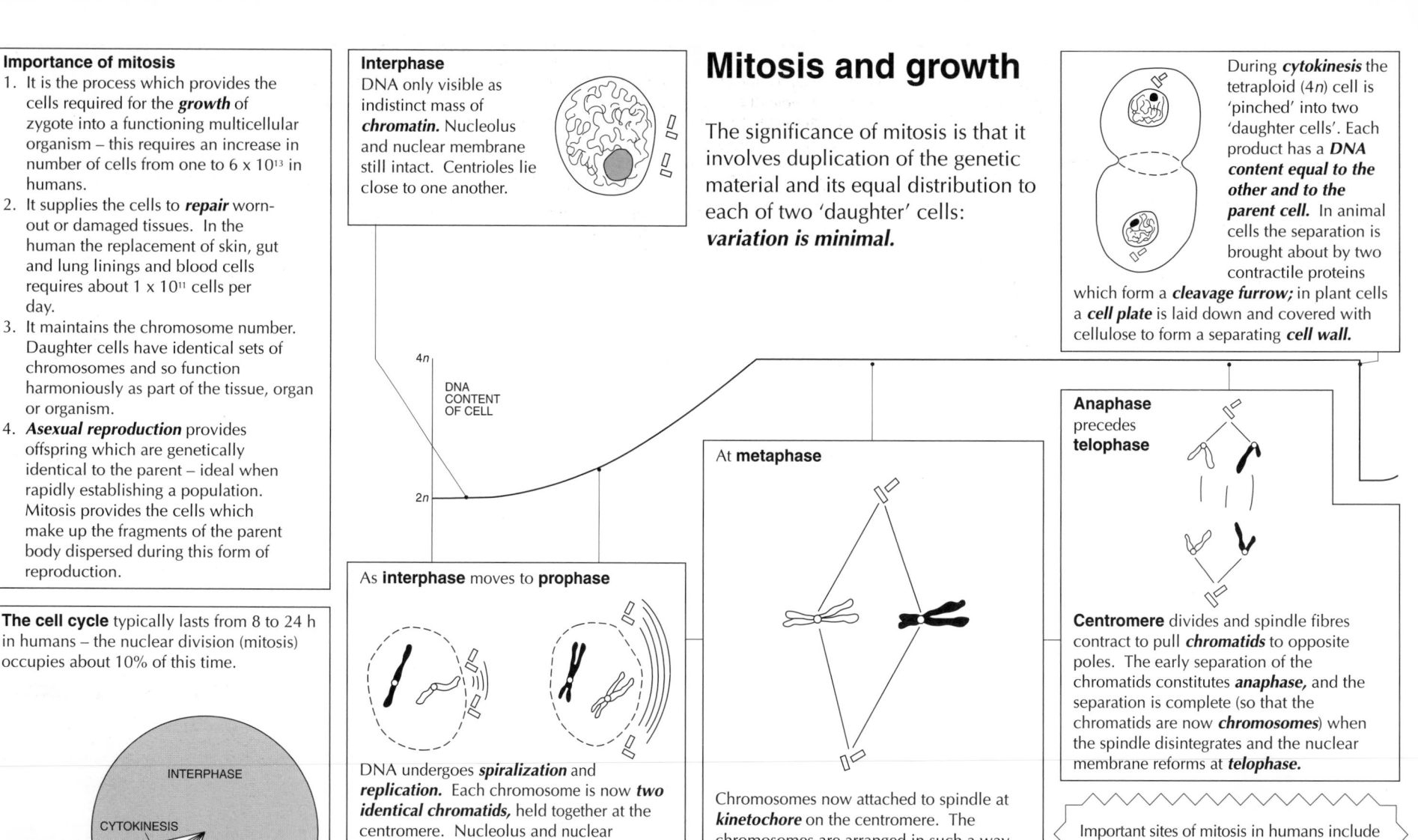

DNA CONTENT OF CELL

$4n$

$2n$

INTERPHASE

CYTOKINESIS

CELL DIVISION

TELOPHASE
ANAPHASE
METAPHASE
PROPHASE

NUCLEAR DIVISION (MITOSIS)

Meiosis and variation

Meiosis separates chromosomes, halving the diploid number, and introduces variation to the haploid products.

During **prophase I** each replicated chromosome (comprising two chromatids) pairs with its **homologous partner,** i.e. the diploid number of chromosomes produces the haploid number of homologous pairs.

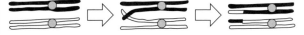

Crossing over occurs when all four chromatids are at **synapsis** (exactly aligned) – non-sister chromatids may cross over, break and reassemble so that **parental** gene combinations are replaced by **recombinants.** This is a major source of **genetic variation.**

Importance of meiosis

1. It must occur in sexually reproducing organisms or the chromosome number would be doubled at fertilization.

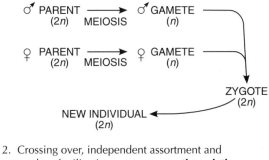

2. Crossing over, independent assortment and random fertilization promote **genetic variation.** This provides new material for natural selection to work on during evolution.

At **anaphase I** and **telophase I** there is separation of **whole chromosomes** (i.e. of **pairs** of chromatids).

The products of meiosis I now contain the **haploid** (*n*) number of chromosomes, although each chromosome comprises two chromatids.

During the **second meiotic division** (metaphase II, anaphase II and telophase II) there is a modified mitosis which **separates the two sister chromatids of each chromosome.**

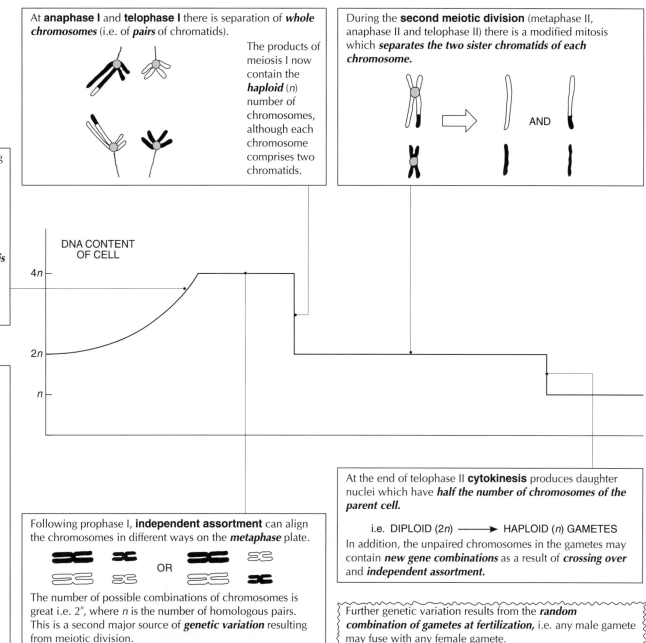

DNA CONTENT OF CELL

$4n$

$2n$

n

Following prophase I, **independent assortment** can align the chromosomes in different ways on the **metaphase** plate.

OR

The number of possible combinations of chromosomes is great i.e. 2^n, where *n* is the number of homologous pairs. This is a second major source of **genetic variation** resulting from meiotic division.

At the end of telophase II **cytokinesis** produces daughter nuclei which have **half the number of chromosomes of the parent cell.**

i.e. DIPLOID ($2n$) ⟶ HAPLOID (*n*) GAMETES

In addition, the unpaired chromosomes in the gametes may contain **new gene combinations** as a result of **crossing over** and **independent assortment.**

Further genetic variation results from the **random combination of gametes at fertilization,** i.e. any male gamete may fuse with any female gamete.

Monohybrid inheritance

Phenylketonuria in humans is an example of *monohybrid inheritance.*

Phenylketonuria results from a lack of the liver enzyme *phenylalanine hydroxylase:* blood phenylalanine levels are raised, causing a number of effects.

PHENYLALANINE in diet

Normal pathway → TYROSINE → MELANIN

Mutant pathway → PHENYLPYRUVATE

Accumulation in PKU sufferers causes mental retardation, abnormal muscle tone and body movement.

Absence in PKU sufferers causes pale hair, skin and eyes.

Early diagnosis advises a phenylalanine-restricted diet for children, markedly reducing PKU symptoms.

If one parent is homozygous normal, the other homozygous mutant

The homozygous normal individual is represented as NN, the mutant individual as nn since in this case normal is dominant to mutant (phenylketonuric).

PARENTAL GENERATION

At meiosis, only one of the two chromosomes (thus only one of the two alleles) can be transmitted to the gamete: *Mendel's First Law.*

GAMETES

1ST FILIAL GENERATION

At *fertilization,* fusion of gametes to form a zygote restores the diploid number.

This individual is *genotypically* heterozygous, but *phenotypically* normal, i.e. a *carrier* of the recessive allele for phenylketonuria.

A *Punnett square* can be used to predict the possible combinations of alleles in the zygote.

	Gametes from father	
Gametes from mother	N	n
N	NN	Nn
n	Nn	nn

If both parents are carriers (i.e. heterozygous)

PARENTAL (P) GENERATION   Nn × Nn

GAMETES

1ST FILIAL (F₁) GENERATION

NN — Phenotypically and genotypically normal

Nn — Phenotypically normal but a carrier

nn — Phenotypically and genotypically recessive

3 NORMAL : 1 MUTANT

N.B. the 3 : 1 ratio is only approximate unless the number of offspring is very large (unlikely in humans), because

1. Alleles may not be distributed between viable gametes in equal numbers.

2. Fusion of gametes is completely random – it is a matter of chance whether one male gamete fuses with a particular female gamete.

Other significant examples of monohybrid inheritance in humans

Albinism: autosomal recessive

Cystic fibrosis: autosomal recessive (the most common lethal allele in Caucasian populations)

Huntington's disease: autosomal dominant

Tay-Sach's disease: autosomal recessive (prevalent in Jews of Eastern European origin)

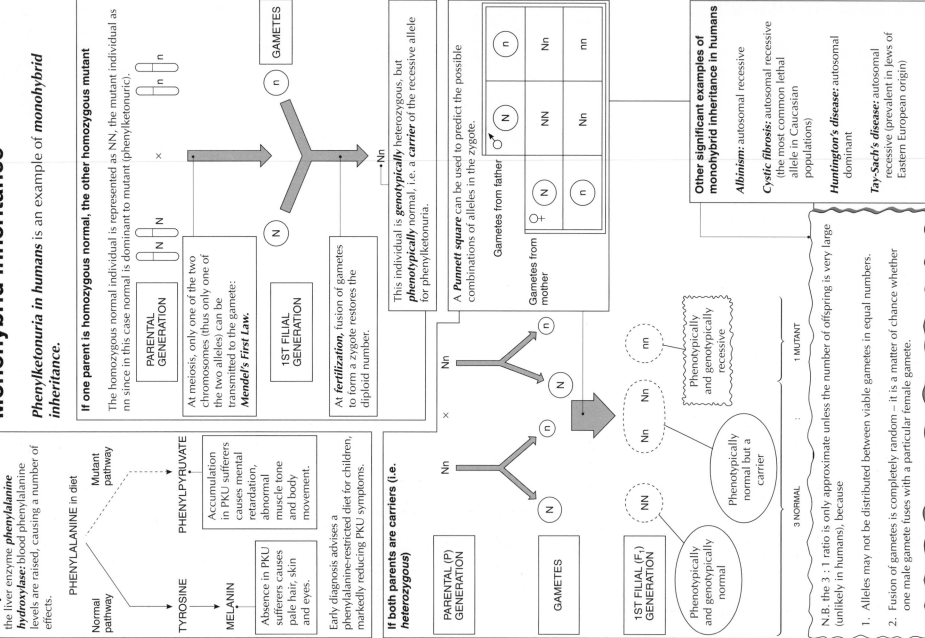

Dihybrid inheritance

the transmission of **two pairs of alleles** at the same time but independently of one another.

Gregor Mendel was an Austrian monk who studied patterns of inheritance in the garden pea. He made several proposals concerning these patterns: these proposals have been found to hold true for many other organisms. One organism which has been widely used for experiments on inheritance is the fruit fly, *Drosophila melanogaster*, which has a number of advantages for such studies:

1. It has a rapid generation time (10 days) and produces many offspring from each mating so that statistical analysis can be applied to results.

2. It is easily cultured in small, convenient containers (milk bottles!) on a simple growth medium.

3. Males and females are easily distinguished so that controlled matings are possible.

4. The flies have a number of obvious external characteristics which are easily mutated. These include wing length, body colour and eye shape.

Flies which differ in **two pairs of characteristics** (e.g. **body colour** and **wing shape**) may be mated.

Parental and filial generations

PARENTAL (P) GENERATION	Grey body, long wing ×	Black body, vestigial wing
1ST FILIAL (F₁) GENERATION	All have grey body, long wing	

Self-fertilization between F₁ individuals

2ND FILIAL (F₂) GENERATION	
Grey body, long wing	9
Grey body, vestigial wing	3
Black body, long wing	3
Black body, vestigial wing	1

A phenotypic ratio of 9:3:3:1 would seem to be complex, but Mendel explained this as **two separate monohybrid crosses** (i.e. **grey v. black** and **long v. vestigial) occurring at the same time.**

ie.

GREY v. BLACK

= (9+3) : (3+1)

= 12 : 4

= 3 : 1

LONG v. VESTIGIAL

= (9+3) : (3+1)

= 12 : 4

= 3 : 1

i.e. 9 : 3 : 3 : 1 is the same as 3 : 1 × 3 : 1

Thus the inheritance of **body colour** had not influenced the inheritance of **wing shape.**

Mendel's Second Law (the law of independent assortment)

'Each member of a pair of alleles may combine randomly with either of another pair'

In this example, the allele for **grey** body may combine equally often with the allele for **long** wing or with the allele for **vestigial** wing.

Using genetic symbols

Let G = grey, g = black, L = long, l = vestigial

P	GGLL	×	ggll

Gametes

(GL) (gl)

F₁

GgLl

Gametes

(GL) (Gl) (gL) (gl)

These will be produced in equal numbers, according to Mendel's Second Law.

F₂

The possible combinations of gametes are most easily derived using a **Punnett square.**

♂ GAMETES ♀ GAMETES	GL	Gl	gL	gl
GL	GGLL	GGLl	GgLL	GgLl
Gl	GGLl	GGll	GgLl	Ggll
gL	GgLL	GgLl	ggLL	ggLl
gl	GgLl	Ggll	ggLl	ggll

or, phenotypically

- 9 GREY BODY, LONG WING (both G and L in zygote)
- 3 GREY BODY, VESTIGIAL WING (G and ll in zygote)
- 3 BLACK BODY, LONG WING (gg and L in zygote)
- 1 BLACK BODY, VESTIGIAL WING (ggll in zygote)

Sex linkage and the inheritance of sex

A KARYOTYPE is obtained by rearranging photographs of stained chromosomes observed during mitosis. Such a karyotype indicates that

1 the chromosomes are arranged in homologous pairs. In humans there are 23 pairs and we say that the **diploid number** is 46 (2n = 46 = 2 × 23).

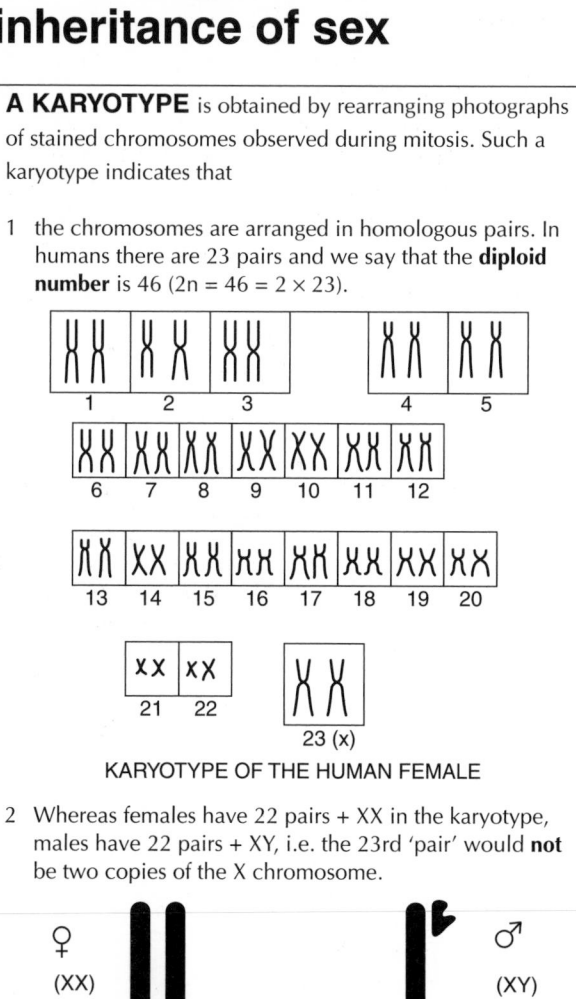

KARYOTYPE OF THE HUMAN FEMALE

2 Whereas females have 22 pairs + XX in the karyotype, males have 22 pairs + XY, i.e. the 23rd 'pair' would **not** be two copies of the X chromosome.

♀ (XX)   ♂ (XY)

The Y chromosome is so small that there is little room for any genes other than those responsible for 'maleness', but the X chromosome can carry some additional genes as well as those for 'femaleness'. These additional genes are **X-linked** (usually described as **sex-linked**).

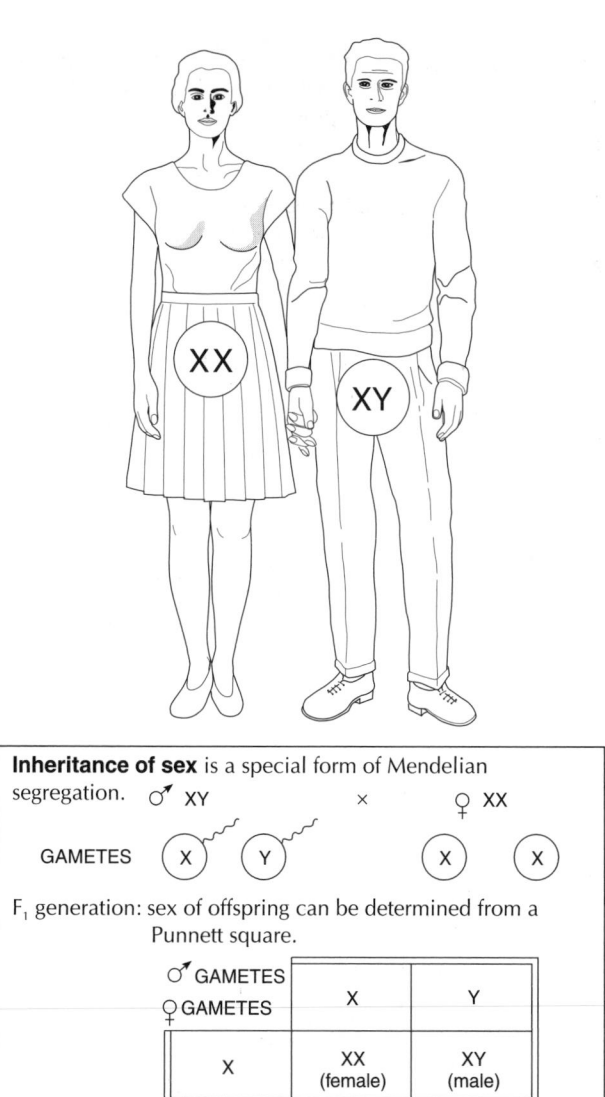

Inheritance of sex is a special form of Mendelian segregation.   ♂ XY   ×   ♀ XX

GAMETES   Ⓧ Ⓨ   Ⓧ Ⓧ

F₁ generation: sex of offspring can be determined from a Punnett square.

♂ GAMETES ♀ GAMETES	X	Y
X	XX (female)	XY (male)
X	XX (female)	XY (male)

Theoretically there should be a 1 : 1 ratio of male : female offspring. In humans various factors can upset the ratio – the Y sperm tend to have greater mobility; the XY zygote and embryo is more delicate than the XX embryo. The balance is just about maintained.

Any genes on the X chromosomes will be inherited by both sexes, but whereas the male can only receive **one** of the alleles (he will be XY, and therefore must be **homozygous** for the X-linked allele) the female will be XX and thus may be either **homozygous** or **heterozygous**. This gives females a tremendous genetic advantage since any recessive lethal allele will not be expressed in the heterozygote.

For example, **haemophilia** is an X-linked condition.

Normal gene = H, mutant gene = h

PARENTS $X^H X^h$ x $X^H Y$
                 female, carrier   male, normal

i.e. **both parents have normal phenotype …**

GAMETES   Ⓧ^{H} Ⓧ^{h} Ⓧ^{H} Ⓨ

F₁ OFFSPRING

♂ GAMETES ♀ GAMETES	X^H	Y
X^H	$X^H X^H$	$X^H Y$
X^h	$X^H X^h$	$X^h Y$

i.e.   normal,   carrier,   normal,   haemophiliac,
       female   female   male   male
       $X^H X^H$ $X^H X^h$ $X^H Y$ $X^h Y$

… but may have a haemophiliac son.

N.B. A haemophiliac daughter ($X^h X^h$) could only be produced if **both** parents contributed X^h i.e. if the father was haemophiliac, in which case his condition would be known and the daughter's condition might be expected.

Another important X-linked condition is **red-green colour blindness**; thus many more males than females cannot distinguish red from green.

Artificial selection occurs when humans, rather than environmental factors, determine which genotypes will pass to successive generations.

POLYPLOIDY AND PLANT BREEDING

Polyploids contain *multiple sets of chromosomes* (chromosome multiplication can be induced by treatment with *colchicine* during mitosis – this inhibits spindle formation and prevents chromatid separation).

Autopolyploids (all chromosomes from the *same* species) e.g. all *bananas* are *triploid* – they are infertile and contain no seeds. Most *potatoes* are *tetraploid* – cells are bigger and tubers are larger. Cultivated *strawberries* are *octoploid*.

Allopolyploidy (sets of chromosomes from *different* species) is possible if the two species have a chromosome complement similar in number and shape. This might allow plant breeders to *combine the beneficial characteristics of more than one species.*

The evolution of *bread wheat* is an important example.

Wild wheat: has brittle ears which fall off on harvesting.
↓ SELECTIVE BREEDING
Einkorn wheat: non-brittle but low yielding.
↓ POLYPLOIDY with *Agropyron* grass
Emmer wheat: high yielding but difficult to separate seed during threshing.
↓ POLYPLOIDY with *Aegilops* grass
Bread wheat: high yielding with easily separated 'naked' seeds.

INBREEDING AND OUTBREEDING

Outbreeding occurs when there is selective controlled reproduction between members of genetically distant populations (different strains or even, for plants, closely related but different species).

Inbreeding occurs when there is selective reproduction between closely related individuals, e.g. between offspring of same litter or between parent and child.

Tends to *introduce new and superior phenotypes:* the progeny are known as *hybrids* and the development of improved characteristics is called *heterosis* or *hybrid vigour.*

e.g. introduction of disease resistance from wild sheep to domestic strains;
combination of shorter-stemmed 'wild' wheat and heavy yielding 'domestic' wheat.

This may result from increased numbers of dominant alleles or from new opportunities for gene interaction.

Genetic conservation i.e. the retention of 'genes' in the form of seed banks or rare breed collections is vital for outbreeding.

Tends to *maintain desirable characteristics*

e.g. uniform height in maize (easier mechanical harvesting); maximum oil content of linseed (more economical extraction); milk production by Jersey cows (high cream content).

But it may cause *reduced fertility* and *lowered disease resistance* as genetic variation is reduced.

Thus inbreeding is not favoured by animal breeders.

PROTOPLAST FUSION is a modern method for production of hybrids in plants.

CELL SUSPENSION OF PLANT A   CELL SUSPENSION OF PLANT B

CELL WALLS REMOVED BY ENZYMES

FUSION OF PROTOPLASTS

HYBRID

e.g. production of virus-resistant tobacco.

TECHNIQUES WITH ANIMALS are less well advanced than those with plants because:

a   animals have a longer generation time and few offspring
b   more food will be made available from improved plants
c   there are many ethical problems which limit genetic experiments with animals.

Two important animal techniques are:

Artificial insemination: allows sperm from a male with desirable characteristics to fertilize a number of female animals.

Embryo transplantation: allows the use of *surrogate mothers* (thus increasing number of offspring) and *cloning* (production of many identical animals with the desired characteristics).

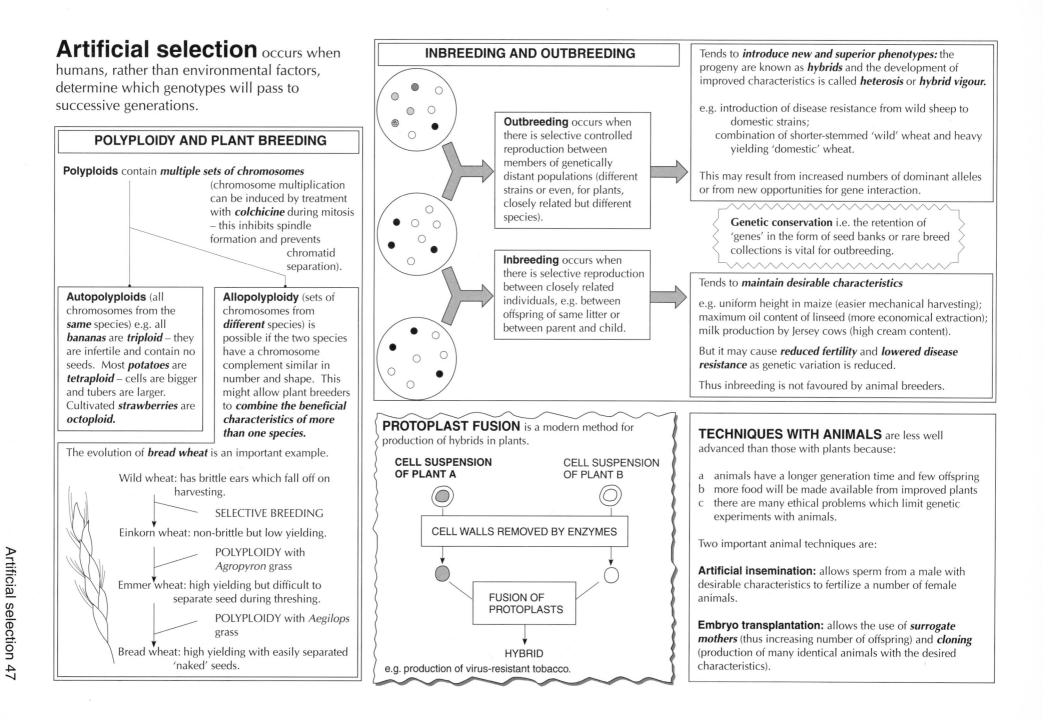

Origins of variation

May be **non-heritable** (e.g. sunburn in a light-skinned individual) or **heritable** (e.g. skin colour in different races). The second type, which result from genetic changes, are the most significant in evolution.

Mutation is any change in the structure or the amount of DNA in an organism.

A **gene** or **point mutation** occurs at a single locus on a chromosome – most commonly by **deletion, addition** or **substitution** of a nucleotide base. Examples are sickle cell anaemia, phenylketonuria and cystic fibrosis.

A **change in chromosome structure** occurs when a substantial portion of a chromosome is altered. For example, Cri-du-chat syndrome results from **deletion** of a part of human chromosome 5, and a form of white blood cell cancer follows **translocation** of a portion of chromosome 8 to chromosome 14.

Aneuploidy (typically the **loss or gain of a single chromosome**) results from **non-disjunction** in which chromosomes fail to separate at anaphase of meiosis. The best known examples are Down's syndrome (extra chromosome 21), Klinefelter's syndrome (male with extra X chromosome) and Turner's syndrome (female with one fewer X chromosome).

Polyploidy (the presence of additional **whole sets of chromosomes**) most commonly occurs when one or both gametes is diploid, forming a polyploid on fertilization. Polyploidy is rare in animals, but there are many important examples in plants, e.g. bananas are triploid, and tetraploid tomatoes are larger and richer in vitamin C.

Variation is the basis of evolution.

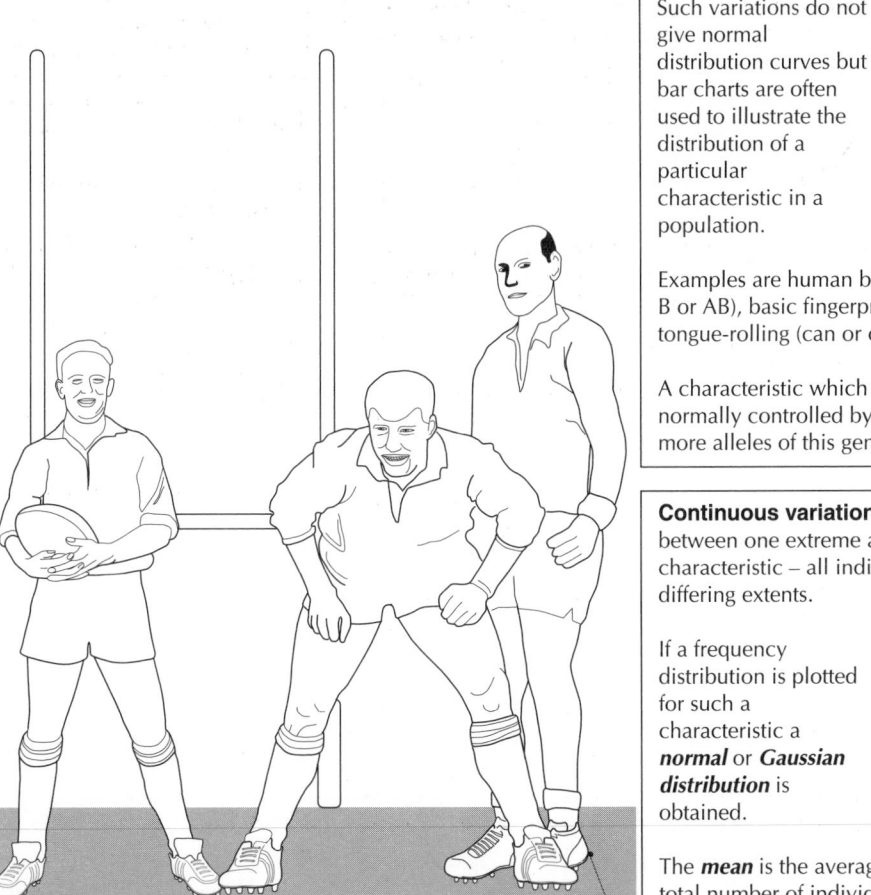

Sexual or genetic recombination is a most potent force in evolution, since it reshuffles genes into new combinations. It may involve

- **free assortment** in gamete formation
- **crossing over** during meiosis
- **random fusion** during zygote formation.

Discontinuous variation occurs when a characteristic is either present or absent (the two extremes) and there are no intermediate forms.

Such variations do not give normal distribution curves but bar charts are often used to illustrate the distribution of a particular characteristic in a population.

Examples are human blood groups in the ABO system (O, A, B or AB), basic fingerprint forms (loop, whorl or arch) and tongue-rolling (can or cannot).

A characteristic which shows discontinuous variation is normally controlled by a single gene – there may be two or more alleles of this gene.

Continuous variation occurs when there is a gradation between one extreme and the other of some given characteristic – all individuals exhibit the characteristic but to differing extents.

If a frequency distribution is plotted for such a characteristic a **normal** or **Gaussian distribution** is obtained.

The **mean** is the average number of such a group (i.e. the total number of individuals divided by the number of groups), the **mode** is the most common of the groups and the **median** is the central value of a set of values.

Typical examples are height, mass, handspan, or number of leaves on a plant.

Characteristics which show continuous variation are controlled by the combined effect of a number of genes, called **polygenes,** and are therefore **polygenic characteristics.**

Natural selection may be a potent force in *evolution*.

Much variation is of the **continuous type**, i.e. a range of phenotypes exists between two extremes. The range of phenotypes within the environment will typically show a **normal distribution.**

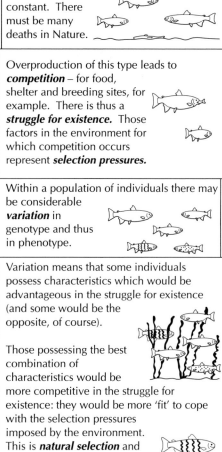

PHENOTYPIC CLASSES

The modern **neo-Darwinian** theory accepts that:

1. Some harmful alleles may survive, but the reproductive potential of the individual possessing such an allele will be reduced.

2. Selective advantages and disadvantages of an allele relate to one environment at one particular time, i.e. an allele does not always contribute to 'fitness' but only under certain conditions.

Plants and animals in Nature produce more offspring than can possibly survive, yet the population remains relatively constant. There must be many deaths in Nature.

Overproduction of this type leads to **competition** – for food, shelter and breeding sites, for example. There is thus a **struggle for existence.** Those factors in the environment for which competition occurs represent **selection pressures.**

Within a population of individuals there may be considerable **variation** in genotype and thus in phenotype.

Variation means that some individuals possess characteristics which would be advantageous in the struggle for existence (and some would be the opposite, of course).

Those possessing the best combination of characteristics would be more competitive in the struggle for existence: they would be more 'fit' to cope with the selection pressures imposed by the environment. This is **natural selection** and promotes **survival of the fittest.**

If variation is **heritable** (i.e. caused by an alteration in genotype) new generations will tend to contain a higher proportion of individuals suited to survival.

Stabilizing selection favours intermediate phenotypic classes and operates against extreme forms – there is thus a **decrease** in the frequency of alleles representing the extreme forms.

Stabilizing selection operates when the phenotype corresponds with optimal environmental conditions, and competition is not severe. It is probable that this form of selection has favoured heterozygotes for **sickle cell anaemia** in an environment in which **malaria** is common, and also works against **extremes of birth weight** in humans.

Directional selection favours one phenotype at one extreme of the range of variation. It moves the phenotype towards a new optimum environment; then stabilizing selection takes over. There is a change in the allele frequencies corresponding to the new phenotype.

Directional selection has occurred in the case of the peppered moth, *Biston betularia*, where the dark form was favoured in the sooty suburban environments of Britain during the industrial revolution: **industrial melanism.** Another significant example is the development of **antibiotic resistance** in populations of bacteria – mutant genes confer an advantage in the presence of an antibiotic – and pesticide resistance in agricultural pests.

Disruptive selection is the rarest form of selection and is associated with a variety of selection pressures operating within one environment.

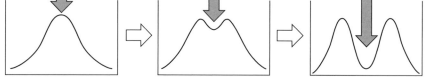

This form of selection promotes the co-existence of more than one phenotype, the condition of **polymorphism (balanced** polymorphism when no one selective agent is more important than any other). Important examples are:
1. **Colour** (yellow/brown) and **banding pattern** (from 0–5) in *Cepaea nemoralis.*
2. **Three phenotypes** (corresponding to HbHb/HbHb^s/Hb^{s}Hb^s) show an uneven distribution of the sickle-cell allele in different areas of the world.

Reproductive isolation and speciation

Reproductive isolation is essential for speciation: ***allopatric speciation*** occurs when populations occupy different environments; ***sympatric speciation*** occurs when populations are reproductively isolated within the same environment.

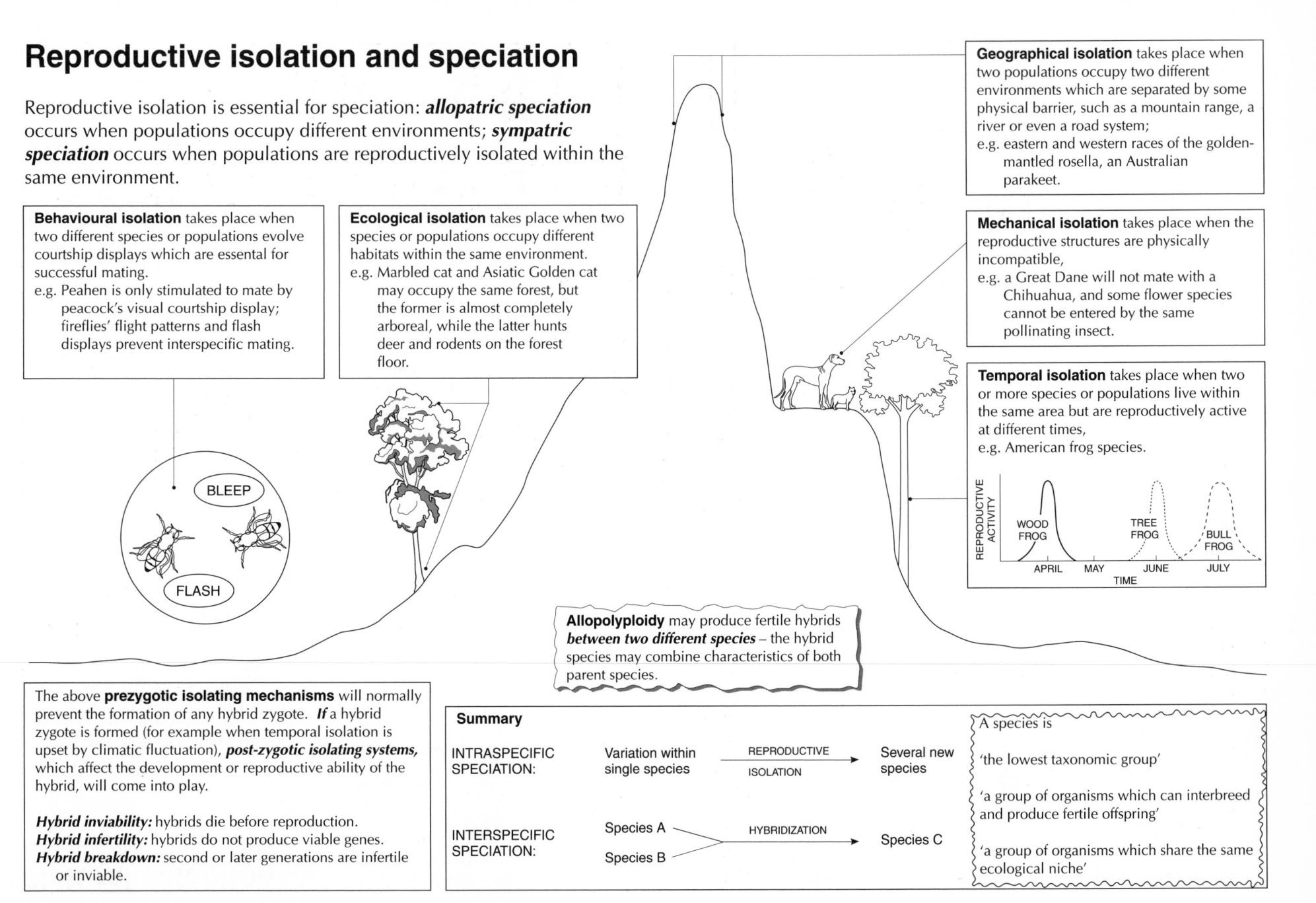

Behavioural isolation takes place when two different species or populations evolve courtship displays which are essential for successful mating.
e.g. Peahen is only stimulated to mate by peacock's visual courtship display; fireflies' flight patterns and flash displays prevent interspecific mating.

BLEEP

FLASH

Ecological isolation takes place when two species or populations occupy different habitats within the same environment.
e.g. Marbled cat and Asiatic Golden cat may occupy the same forest, but the former is almost completely arboreal, while the latter hunts deer and rodents on the forest floor.

Geographical isolation takes place when two populations occupy two different environments which are separated by some physical barrier, such as a mountain range, a river or even a road system;
e.g. eastern and western races of the golden-mantled rosella, an Australian parakeet.

Mechanical isolation takes place when the reproductive structures are physically incompatible,
e.g. a Great Dane will not mate with a Chihuahua, and some flower species cannot be entered by the same pollinating insect.

Temporal isolation takes place when two or more species or populations live within the same area but are reproductively active at different times,
e.g. American frog species.

REPRODUCTIVE ACTIVITY

WOOD FROG   TREE FROG   BULL FROG

APRIL   MAY   JUNE   JULY
TIME

Allopolyploidy may produce fertile hybrids ***between two different species*** – the hybrid species may combine characteristics of both parent species.

The above **prezygotic isolating mechanisms** will normally prevent the formation of any hybrid zygote. **If** a hybrid zygote is formed (for example when temporal isolation is upset by climatic fluctuation), ***post-zygotic isolating systems,*** which affect the development or reproductive ability of the hybrid, will come into play.

Hybrid inviability: hybrids die before reproduction.
Hybrid infertility: hybrids do not produce viable genes.
Hybrid breakdown: second or later generations are infertile or inviable.

Summary

INTRASPECIFIC SPECIATION:	Variation within single species	REPRODUCTIVE ISOLATION →	Several new species
INTERSPECIFIC SPECIATION:	Species A / Species B	HYBRIDIZATION →	Species C

A species is

'the lowest taxonomic group'

'a group of organisms which can interbreed and produce fertile offspring'

'a group of organisms which share the same ecological niche'

Naming and classifying living organisms

THE BINOMIAL SYSTEM OF NOMENCLATURE

provided by Linnaeus

successful since
(a) each species has a unique name
(b) shows which species are closely related e.g. *Panthera leo* and *Panthera tigris*

incorporates both species and genus into specific name
e.g. *Homo sapiens*
(i) Underline or *italicise*
(ii) Capital letter for genus, lower case for species.

the 'naming' of organisms (usually after they have been placed into 'groups' or taxons)

needed to aid communication between groups of scientists so 'name' must be unambiguous and easily understood

usually in Latin since
(a) original scientific language
(b) universally accepted

HIERARCHICAL CLASSIFICATION OF THE LION

On moving down the hierarchy of groups, note that there are …

MORE SIMILARITIES
and
FEWER DIFFERENCES
between the members

SPECIES *Panthera leo* — Other species

GENUS (*Panthera*) — Other genera

FAMILY (Felidae) — Other families

ORDER (Carnivora) — Other orders

CLASS (Mammalia) — Other classes

PHYLUM (Chordata) — Other phyla

KINGDOM (Animalia)

All lions can mate and produce fertile offspring with other lions

All *Panthera* (big cats) can roar but cannot purr

All felidae have retractable claws

All carnivores have well-developed carnassial (flesh-cutting) teeth

All mammals have fur and mammary glands

All chordates have a notochord (→backbone)

All animals are ingestive heterotrophs

Look! An ingestive heterotroph with a backbone, mammary glands, carnassial teeth, retractable claws which is about to …

RRROARR!

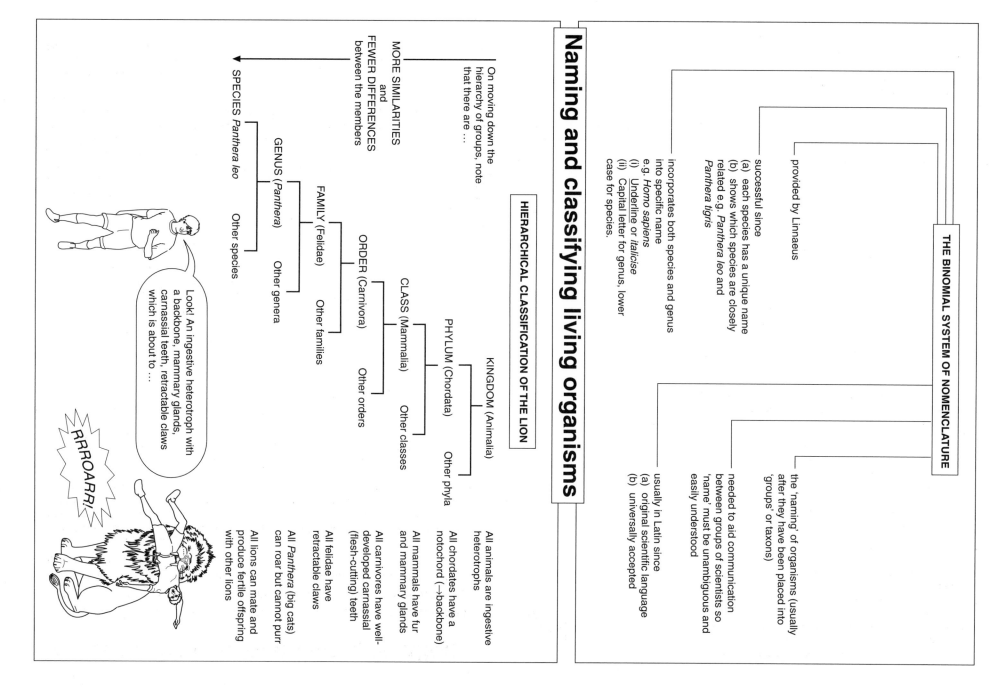

Keys and classification

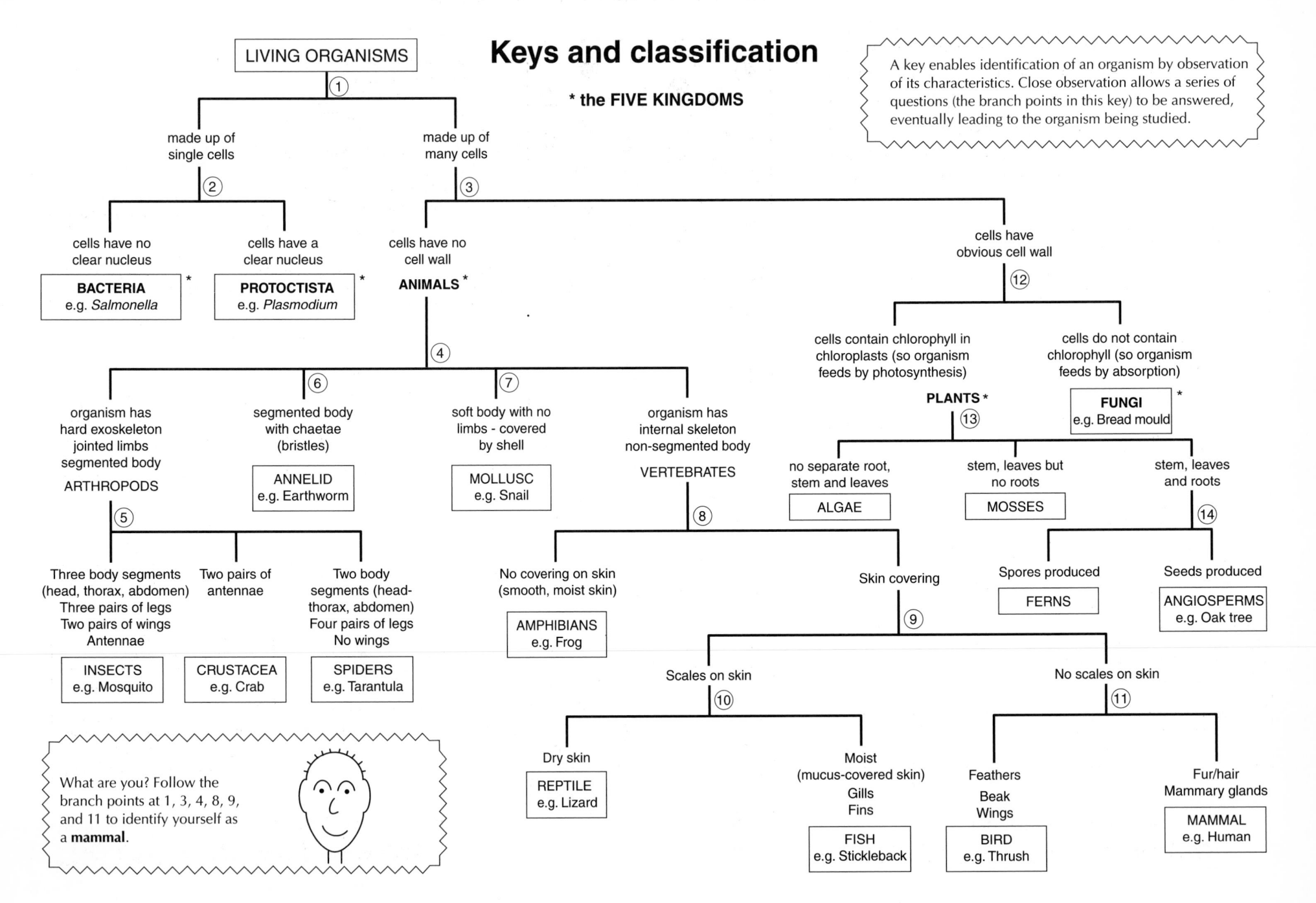

The Chi-squared (χ^2) test:

Analysis of data consisting of discrete (discontinuous) variables

The *expected* ratio of individuals which **show** the dark allele 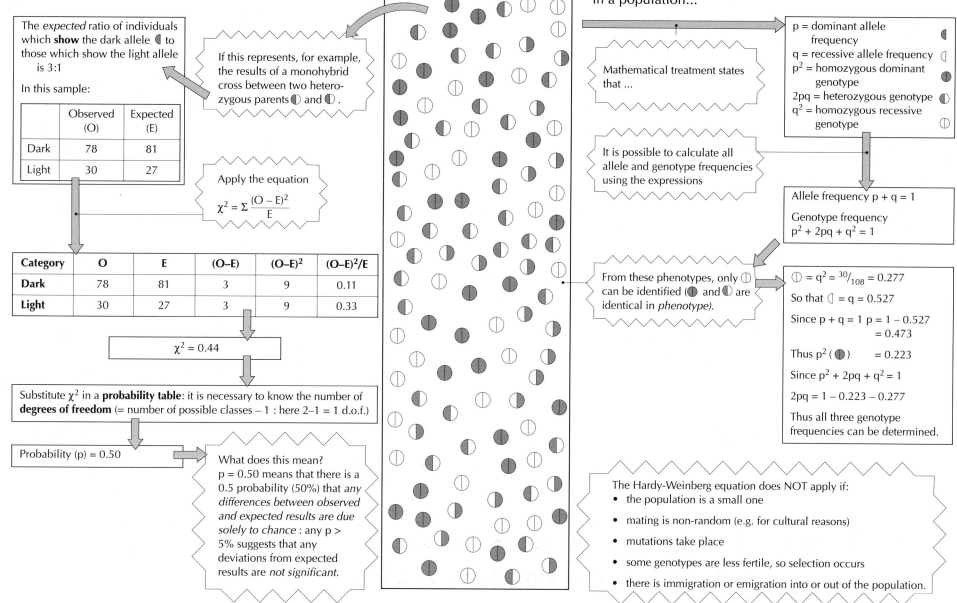 to those which show the light allele is 3:1

In this sample:

	Observed (O)	Expected (E)
Dark	78	81
Light	30	27

If this represents, for example, the results of a monohybrid cross between two hetero-zygous parents ◑ and ◑ .

Apply the equation

$$\chi^2 = \Sigma \frac{(O - E)^2}{E}$$

Category	O	E	(O–E)	(O–E)^2	(O–E)^2/E
Dark	78	81	3	9	0.11
Light	30	27	3	9	0.33

$\chi^2 = 0.44$

Substitute χ^2 in a **probability table**: it is necessary to know the number of **degrees of freedom** (= number of possible classes – 1 : here 2–1 = 1 d.o.f.)

Probability (p) = 0.50

What does this mean?
p = 0.50 means that there is a 0.5 probability (50%) that *any differences between observed and expected results are due solely to chance* : any p > 5% suggests that any deviations from expected results are *not significant*.

The Hardy-Weinberg Equation

Calculation of allele and genotype frequencies in a population...

Mathematical treatment states that ...

p = dominant allele frequency
q = recessive allele frequency
p^2 = homozygous dominant genotype
2pq = heterozygous genotype
q^2 = homozygous recessive genotype

It is possible to calculate all allele and genotype frequencies using the expressions

Allele frequency p + q = 1

Genotype frequency
p^2 + 2pq + q^2 = 1

From these phenotypes, only ◐ can be identified (◑ and ◑ are identical in *phenotype*).

◐ = q^2 = $^{30}/_{108}$ = 0.277

So that ◖ = q = 0.527

Since p + q = 1 p = 1 – 0.527
                      = 0.473

Thus p^2 (●)   = 0.223

Since p^2 + 2pq + q^2 = 1

2pq = 1 – 0.223 – 0.277

Thus all three genotype frequencies can be determined.

The Hardy-Weinberg equation does NOT apply if:
- the population is a small one
- mating is non-random (e.g. for cultural reasons)
- mutations take place
- some genotypes are less fertile, so selection occurs
- there is immigration or emigration into or out of the population.

Calculation of allele and genotype frequencies in a population

Estimating populations

To study the dynamics of a population, or how the distribution of the members of a population is influenced by a biotic or an abiotic factor, it is necessary to estimate the population size. In other words, it will be necessary to count the number of individuals in a population. Such counting is usually carried out by taking **samples** (in which the organisms are in the same proportion as in the whole population) because:

1. counting the whole population would be extremely laborious and time-consuming
2. counting the whole population might cause unacceptable levels of damage to the habitat, or to the population being studied.

The samples must be representative. They should be:

1. of the same size (e.g. a 0.25 m² area of grassland)
2. randomly selected – for example, samples may be taken at predetermined points on an imaginary grid laid over the sampling area. The coordinates of the points may be selected using random numbers generated by a calculator
3. non-overlapping.

QUADRAT SAMPLING

Quandrats are sampling units of a known area. They are most often square, and are usually constructed of wood or metal. The quadrat can be used in simple form, or it may have wire subdivisions to produce a number of sampling points.

Reliable sampling with quadrats requires answers to three questions:
1. What size of quadrat should be used?
2. How many quadrats should be used?
3. Where should the quadrats be positioned?

WHAT SIZE QUADRAT?

If individuals within a population are truly randomly dispersed, then any quadrat size should be equally efficient in the estimation of that population. However, environmental factors are rarely evenly distributed so that the living organisms dependent on them tend to occur in an aggregated distribution. Small quadrats are more efficient in estimating populations (more can be taken, and they can cover a wider range of habitat than larger ones) but there are practical considerations to be taken into account (a small quadrat might not include a dominant tree in a woodland). Optimum quadrat size is determined by counting the number of different species present in quadrats of increasing size.

The optimum quadrant size – 1% increase in quadrat size produce no more than a 0.5% increase in the number of species present

(graph: number of species vs QUADRAT SIZE m²)

RANDOM POSITIONING OF QUADRATS

The position can be chosen using random coordinates, as described earlier.

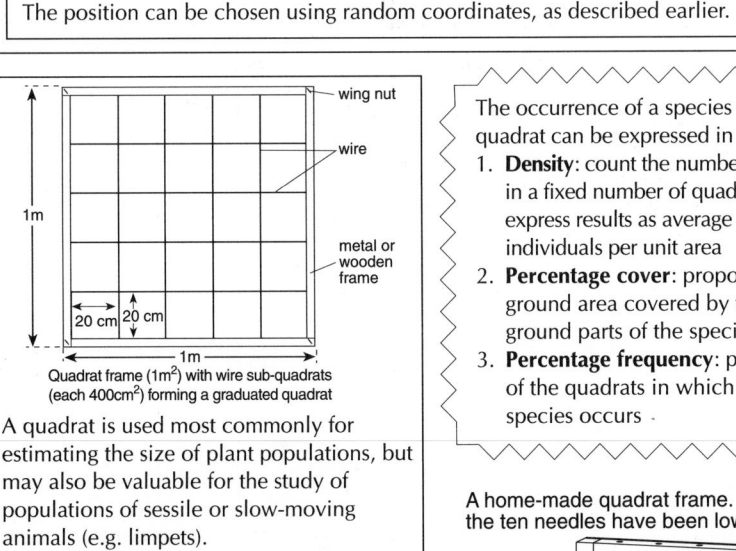

Quadrat frame (1m²) with wire sub-quadrats (each 400cm²) forming a graduated quadrat

A quadrat is used most commonly for estimating the size of plant populations, but may also be valuable for the study of populations of sessile or slow-moving animals (e.g. limpets).

A POINT QUADRAT is used for sampling plant populations in short grassland. It consists of pointed needles pushed through a horizontal wooden or metal frame, usually in groups of ten. Each plant touched by the point of a needle is recorded.

The occurrence of a species within a quadrat can be expressed in several ways:
1. **Density**: count the number of individuals in a fixed number of quadrats, and express results as average number of individuals per unit area
2. **Percentage cover**: proportion of the ground area covered by the above ground parts of the species
3. **Percentage frequency**: percentage of the quadrats in which the species occurs

HOW MANY QUADRATS?

Too few might be unrepresentative, and too many might be tedious and time-consuming. To determine the optimum number, a series of quadrats of the optimum size is placed randomly across the sampling area – the cumulative number of species is recorded after each increase in quadrat number.

(graph: number of species vs NUMBER OF QUADRATS)

TRANSECTS are used to describe the distribution of species in a straight line across a habitat. Transects are particularly useful for describing zonation of species, for example around field or pond margins, or across a marsh. A simple line transect records all of the species which actually touch the rope or tape stretched across the habitat, a belt transect records all of those species present between two lines (perhaps 0.5m² apart), and an interrupted belt transect records all of those species present in a number of quadrats placed at fixed points along a line stretched across the habitat.

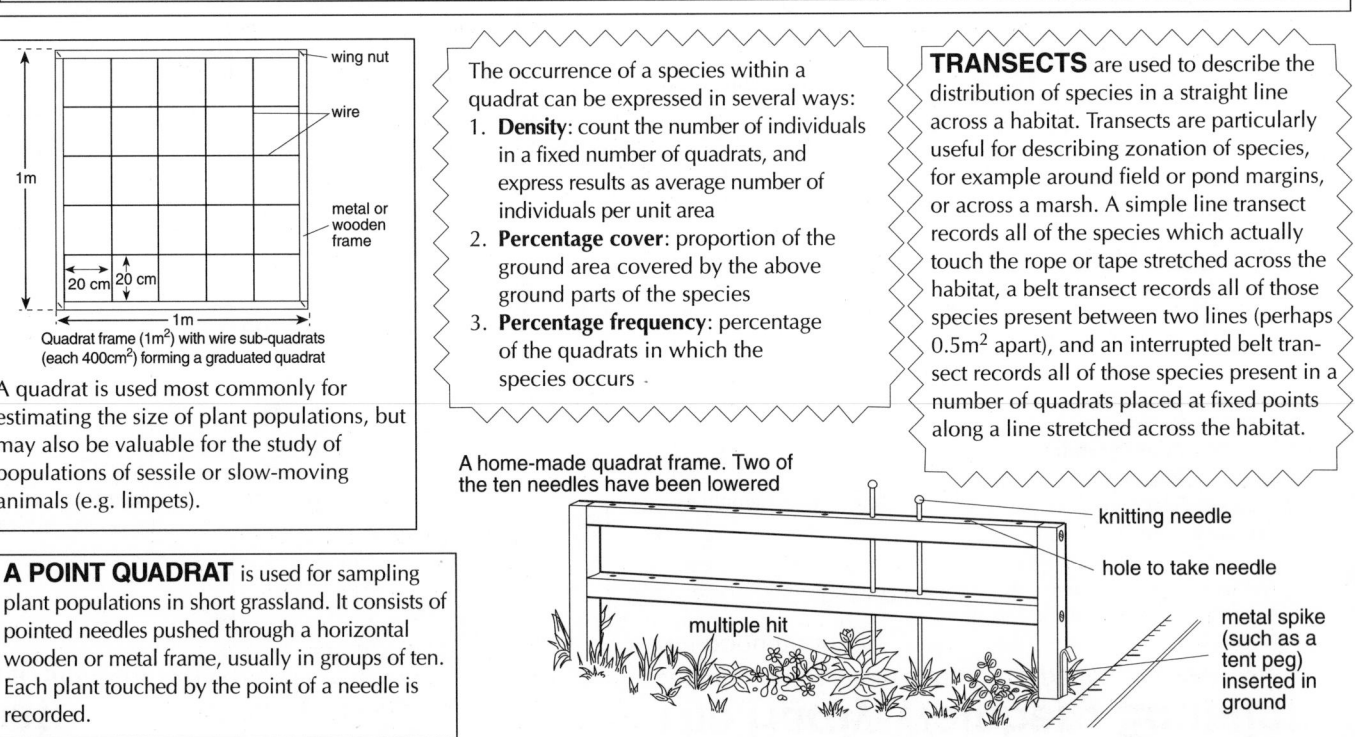

A home-made quadrat frame. Two of the ten needles have been lowered

knitting needle

hole to take needle

multiple hit

metal spike (such as a tent peg) inserted in ground

Sampling motile species

Quadrats and line transects are ideal methods for estimating populations of plants or sedentary animals. Motile animals, however, must be captured before their populations can be estimated. Once more, a representative sample of the population will be counted and the total population estimated from the sample. One important technique is the **mark-recapture** (also known as mark-release-recapture) **method.** This method involves:

1. capturing the organism

2. marking in some way which causes no harm (e.g. beetles can be marked with a drop of waterproof paint on their wing cases, and mice may have a small mark clipped into their fur)

3. releasing the organism to rejoin its population

4. a second sample group from this population is captured and counted at a later date

5. the population size is estimated using the Lincoln Index:

$$\text{Population size} = \frac{n_1 \times n_2}{n_m}$$

where n_1 is the number of individuals marked and released ('1' because it was the first sample), n_2 is the number of individuals caught the second time round ('2' because it was the second sample) and n_m is the number of marked individuals in the second sample ('m' standing for marked).

This method depends on a number of assumptions – failure of any of these to hold up can lead to poor estimates being made. These assumptions are

a. the marked organisms mix randomly back into the normal population (allow sufficient time for this to occur, bearing in mind the mobility of the species)

b. the marked animals are no different to the unmarked ones – they are no more prone to predation, for example

c. changes in population size due to births, deaths, immigration and emigration are negligible

d. the mark does not wear off or grow out during the sampling period.

TULLGREN FUNNEL:
used to collect small organisms from the air spaces of the soil or from leaf litter. The lamp is a source of heat and dehydration – organisms move to escape from it and fall through the sieve (the mesh is fine enough to retain the soil or litter). The animals slip down the smooth-sided funnel and are immobilised in the alcohol. They may then be removed for identification.

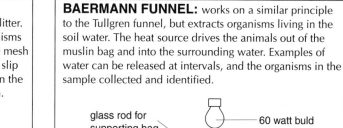

soil sample — 25 watt bulb

16 mesh flour sieve

polythene funnel

80% alcohol

BAERMANN FUNNEL:
works on a similar principle to the Tullgren funnel, but extracts organisms living in the soil water. The heat source drives the animals out of the muslin bag and into the surrounding water. Examples of water can be released at intervals, and the organisms in the sample collected and identified.

glass rod for supporting bag — 60 watt buld

water

soil sample in muslin bag

glass funnel

rubber tubing

clip

beaker

With both Tullgren and Baermann funnels it is essential that samples are treated in identical fashion if results are to be comparative – for example, use fixed sample size, length of exposure to heat source and wattage of lamp.

PITFALL TRAPS:
used to sample arthropods moving over the soil surface.

The roof prevents rainfall from flooding the trap, and also limits access to certain predators. The activities of trapped predators can be prevented by adding a small quantity of methanol to the trap. Bait of meat or ripe fruit can be placed in the trap.

Pitfall traps are often set up on a grid system to investigate the movements of ground animals more systematically.

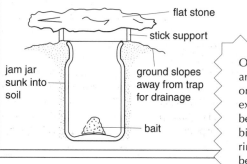

flat stone

stick support

jam jar sunk into soil

ground slopes away from trap for drainage

bait

POOTER:
used to collect specimens of insects and other arthropods which have been extracted from trees or bushes by beating the vegetation over a sheet or tray. Collection in the pooter does not harm the organism, and it can then be returned to its natural habitat.

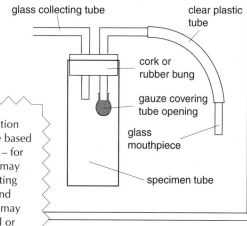

glass collecting tube

clear plastic tube

cork or rubber bung

gauze covering tube opening

glass mouthpiece

specimen tube

Other methods of collection are numerous. Many are based on some form of netting – for example large mist nets may be used to collect migrating birds for identification and ringing, and sweep nets may be used to capture aerial or aquatic arthropods.

Ecology is the 'study of living organisms in relation to their environment'.

A more recent definition is 'the scientific study of the interactions that determine the distribution and abundance of organisms'.

Synecology is the study of groups of organisms associated to form a functional unit of the environment.

Two useful terms are

Community (*biotic community*): all of the populations occupying a given, defined physical area, e.g. all the organisms within a rock pool.

Ecosystem: the biotic community together with the physical (non-living or abiotic) environment, e.g. a rock pool.

Autecology is the study of *single organisms* or *populations of single species* and their relationship to their environment, e.g. Common limpet (*Patella vulgaris*) – an animal of the rocky shore.

What does it *feed on*?

How does it *avoid drying out*?

How does it *minimise damage from wave action*?

What are its *predators*?

How does it *reproduce*?

How do its young *disperse*?

What are its *competitors*?

The **ecological niche** of an organism describes both its location and its function within the community – if two organisms occupy the same ecological niche they are likely to compete with one another.

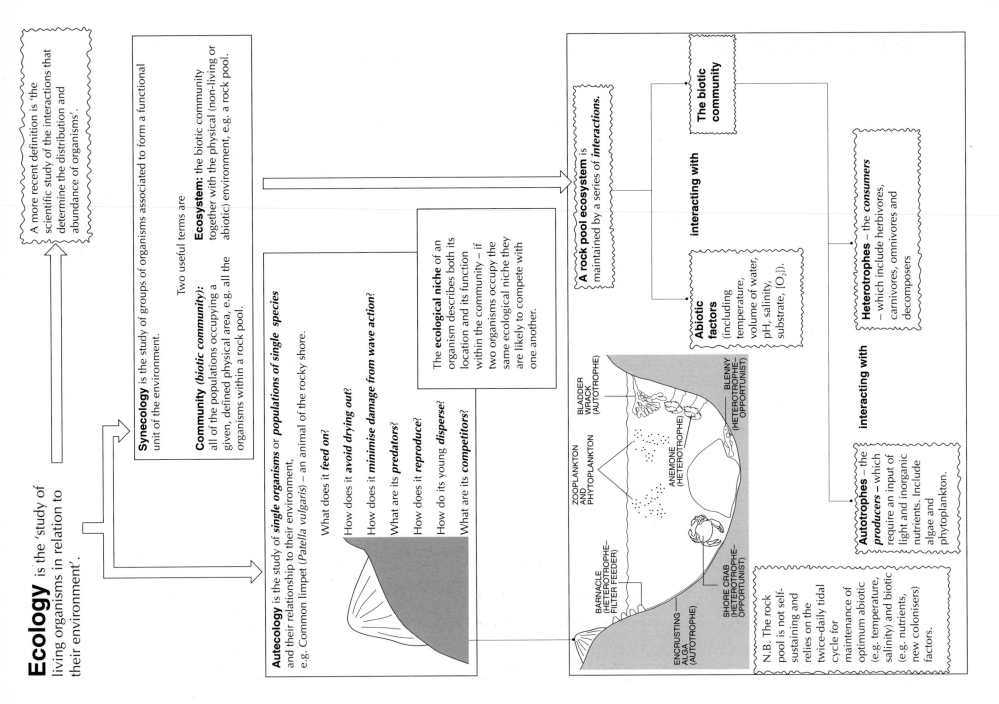

BLADDER WRACK (AUTOTROPHE)

BLENNY (HETEROTROPHE– OPPORTUNIST)

ZOOPLANKTON AND PHYTOPLANKTON

ANEMONE (HETEROTROPHE)

BARNACLE (HETEROTROPHE– FILTER FEEDER)

ENCRUSTING ALGA (AUTOTROPHE)

SHORE CRAB (HETEROTROPHE– OPPORTUNIST)

N.B. The rock pool is not self-sustaining and relies on the twice-daily tidal cycle for maintenance of optimum abiotic (e.g. temperature, salinity) and biotic (e.g. nutrients, new colonisers) factors.

A rock pool ecosystem is maintained by a series of *interactions*.

The biotic community

interacting with

Abiotic factors (including temperature, volume of water, pH, salinity, substrate, [O_2]).

interacting with

Autotrophes – the *producers* – which require an input of light and inorganic nutrients. Include algae and phytoplankton.

Heterotrophes – the *consumers* – which include herbivores, carnivores, omnivores and decomposers.

Law of limiting factors

Photosynthesis is a multi-stage process – for example the Calvin cycle is dependent on the supply of ATP and reducing power from the light reactions – and the principle of limiting factors can be applied.

The rate of a multi-stage process may be subject to different limiting factors at different times. Photosynthesis may be limited by *temperature* during the early part of a summer's day, by *light intensity* during cloudy or overcast conditions or by *carbon dioxide concentration* at other times. The principal limiting factor in Britain during the summer is *carbon dioxide concentration:* the atmospheric $[CO_2]$ is typically only 0.04%. Increased CO_2 emissions from combustion of fossil fuels may stimulate photosynthesis.

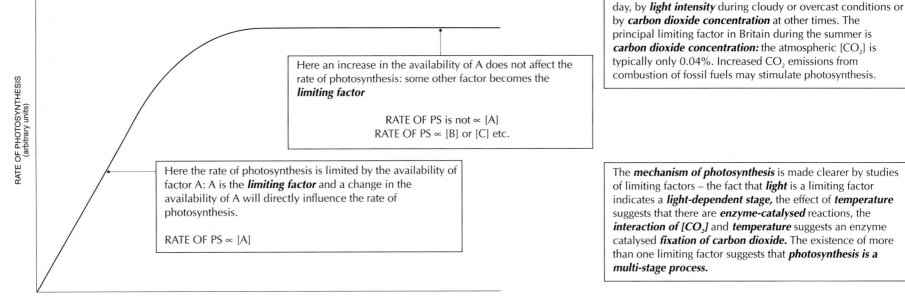

Here an increase in the availability of A does not affect the rate of photosynthesis: some other factor becomes the *limiting factor*

RATE OF PS is not ∝ [A]
RATE OF PS ∝ [B] or [C] etc.

Here the rate of photosynthesis is limited by the availability of factor A: A is the *limiting factor* and a change in the availability of A will directly influence the rate of photosynthesis.

RATE OF PS ∝ [A]

RATE OF PHOTOSYNTHESIS (arbitrary units)

AVAILABILITY OF FACTOR A

The *mechanism of photosynthesis* is made clearer by studies of limiting factors – the fact that *light* is a limiting factor indicates a *light-dependent stage,* the effect of *temperature* suggests that there are *enzyme-catalysed* reactions, the *interaction of $[CO_2]$* and *temperature* suggests an enzyme catalysed *fixation of carbon dioxide.* The existence of more than one limiting factor suggests that *photosynthesis is a multi-stage process.*

The *limiting factors* which affect *photosynthesis* are:

Light intensity: light energy is necessary to generate ATP and $NADPH_2$ during the light dependent stages of photosynthesis.

Carbon dioxide concentration: CO_2 is 'fixed' by reaction with ribulose bisphosphate in the initial reaction of the Calvin cycle.

Temperature: the enzymes catalysing the reactions of the Calvin cycle and some of the light-dependent stages are affected by temperature.

Water availability and **chlorophyll concentration** are not normally limiting factors in photosynthesis.

The study of limiting factors has **commercial and horticultural applications.** Since $[CO_2]$ is a limiting factor, crop production in greenhouses is readily stimulated by raising local carbon dioxide concentrations (from gas cylinders or by burning fossil fuels). It is also clear to horticulturalists that expensive increases in energy consumption for lighting and heating are not economically justified if neither of these is the limiting factor applying under any particular set of conditions.

Factors affecting population growth

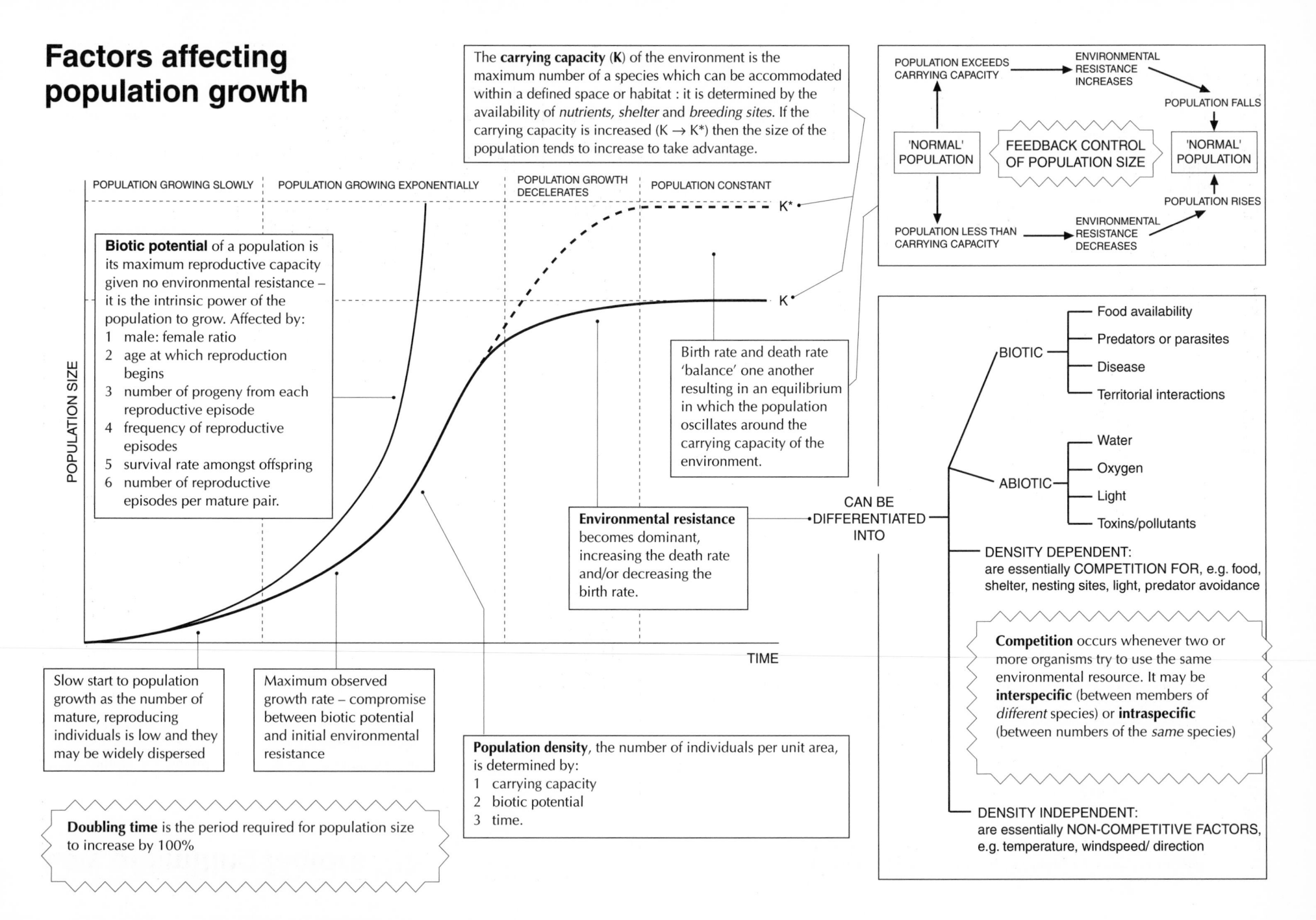

The **carrying capacity** (**K**) of the environment is the maximum number of a species which can be accommodated within a defined space or habitat : it is determined by the availability of *nutrients, shelter* and *breeding sites*. If the carrying capacity is increased (K → K*) then the size of the population tends to increase to take advantage.

FEEDBACK CONTROL OF POPULATION SIZE

POPULATION EXCEEDS CARRYING CAPACITY → ENVIRONMENTAL RESISTANCE INCREASES → POPULATION FALLS

'NORMAL' POPULATION

'NORMAL' POPULATION

POPULATION RISES

POPULATION LESS THAN CARRYING CAPACITY → ENVIRONMENTAL RESISTANCE DECREASES

POPULATION GROWING SLOWLY

POPULATION GROWING EXPONENTIALLY

POPULATION GROWTH DECELERATES

POPULATION CONSTANT

K*

K

POPULATION SIZE

TIME

Biotic potential of a population is its maximum reproductive capacity given no environmental resistance – it is the intrinsic power of the population to grow. Affected by:
1 male: female ratio
2 age at which reproduction begins
3 number of progeny from each reproductive episode
4 frequency of reproductive episodes
5 survival rate amongst offspring
6 number of reproductive episodes per mature pair.

Birth rate and death rate 'balance' one another resulting in an equilibrium in which the population oscillates around the carrying capacity of the environment.

Environmental resistance becomes dominant, increasing the death rate and/or decreasing the birth rate.

CAN BE DIFFERENTIATED INTO

BIOTIC
— Food availability
— Predators or parasites
— Disease
— Territorial interactions

ABIOTIC
— Water
— Oxygen
— Light
— Toxins/pollutants

DENSITY DEPENDENT: are essentially COMPETITION FOR, e.g. food, shelter, nesting sites, light, predator avoidance

Competition occurs whenever two or more organisms try to use the same environmental resource. It may be **interspecific** (between members of *different* species) or **intraspecific** (between numbers of the *same* species)

DENSITY INDEPENDENT: are essentially NON-COMPETITIVE FACTORS, e.g. temperature, windspeed/ direction

Slow start to population growth as the number of mature, reproducing individuals is low and they may be widely dispersed

Maximum observed growth rate – compromise between biotic potential and initial environmental resistance

Population density, the number of individuals per unit area, is determined by:
1 carrying capacity
2 biotic potential
3 time.

Doubling time is the period required for population size to increase by 100%

Human population growth

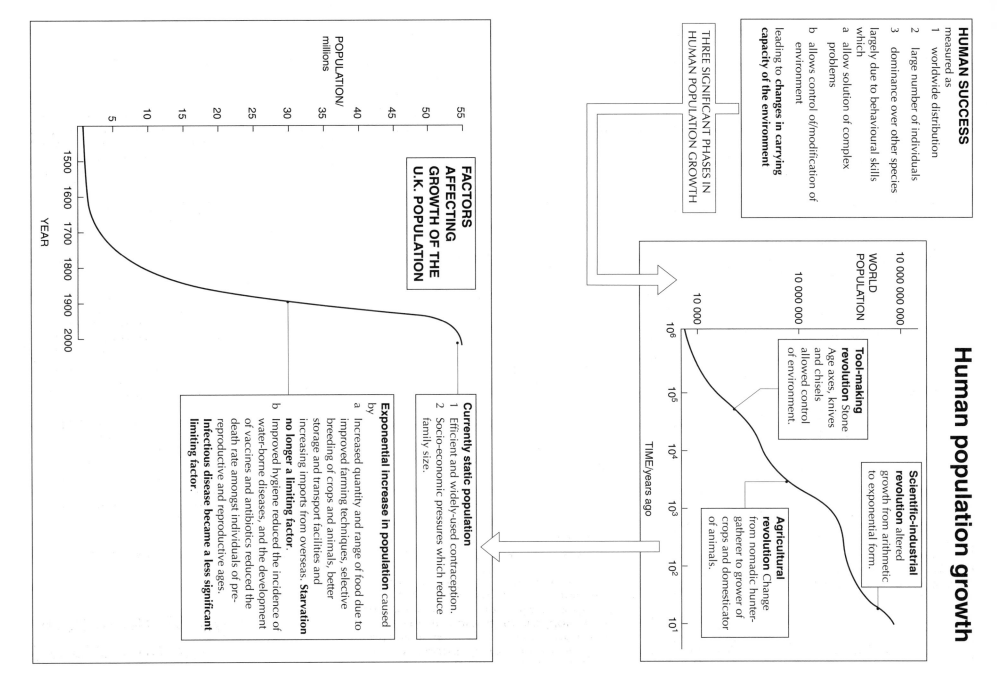

HUMAN SUCCESS

measured as

1 worldwide distribution
2 large number of individuals
3 dominance over other species

largely due to behavioural skills which

a allow solution of complex problems
b allows control of/modification of environment

leading to **changes in carrying capacity of the environment**

THREE SIGNIFICANT PHASES IN HUMAN POPULATION GROWTH

WORLD POPULATION

10 000 000 000
10 000 000
10 000

10^6 10^5 10^4 10^3 10^2 10^1

TIME/years ago

Tool-making revolution Stone Age axes, knives and chisels allowed control of environment.

Scientific-industrial revolution altered growth from arithmetic to exponential form.

Agricultural revolution Change from nomadic hunter-gatherer to grower of crops and domesticator of animals.

FACTORS AFFECTING GROWTH OF THE U.K. POPULATION

POPULATION/ millions

5 10 15 20 25 30 35 40 45 50 55

YEAR

1500 1600 1700 1800 1900 2000

Exponential increase in population caused by

a Increased quantity and range of food due to improved farming techniques, selective breeding of crops and animals, better storage and transport facilities and increasing imports from overseas. **Starvation no longer a limiting factor.**

b Improved hygiene reduced the incidence of water-borne diseases, and the development of vaccines and antibiotics reduced the death rate amongst individuals of pre-reproductive and reproductive ages. **Infectious disease became a less significant limiting factor.**

Currently static population

1 Efficient and widely-used contraception.
2 Socio-economic pressures which reduce family size.

Energy flow through an ecosystem

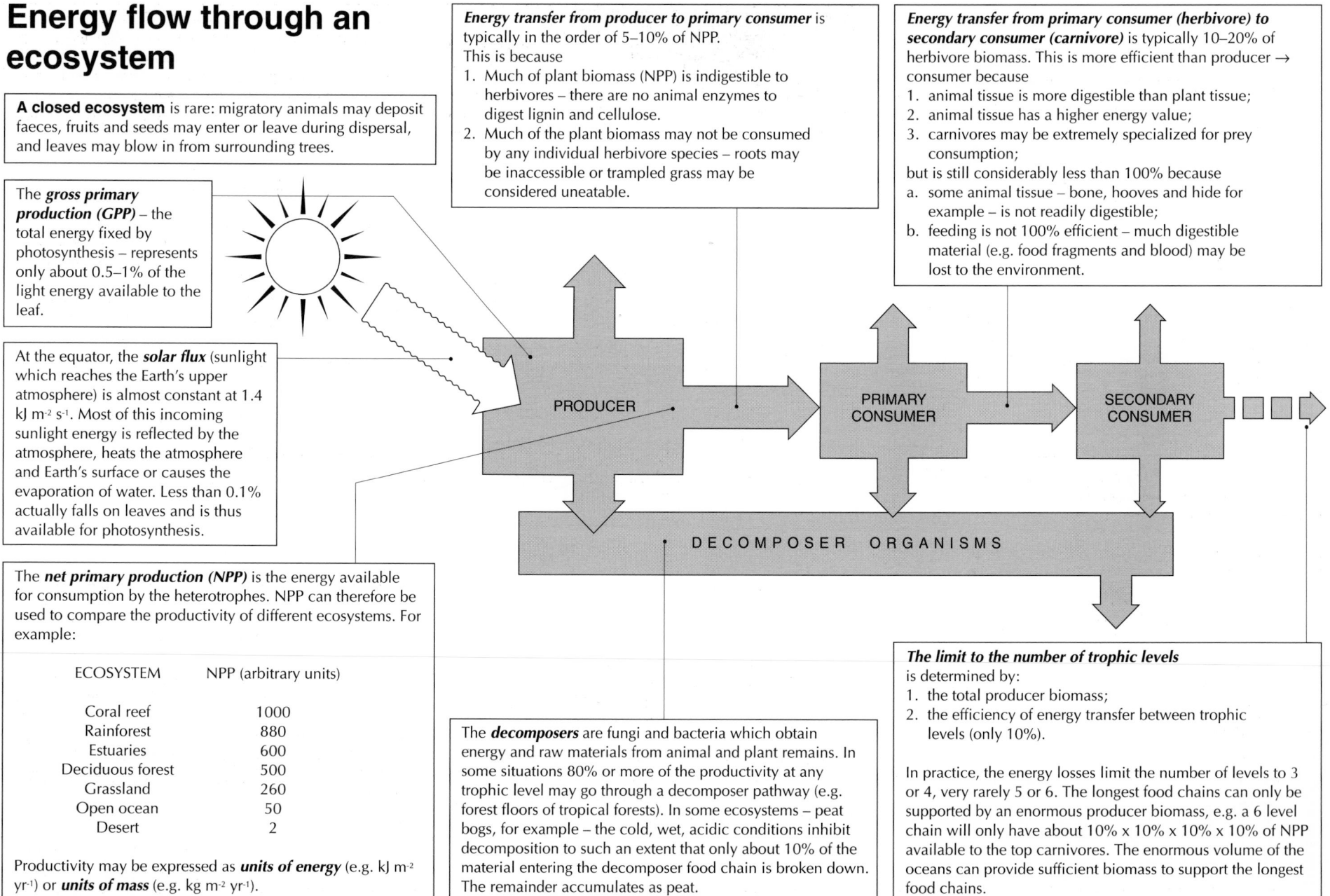

A closed ecosystem is rare: migratory animals may deposit faeces, fruits and seeds may enter or leave during dispersal, and leaves may blow in from surrounding trees.

The **gross primary production (GPP)** – the total energy fixed by photosynthesis – represents only about 0.5–1% of the light energy available to the leaf.

At the equator, the **solar flux** (sunlight which reaches the Earth's upper atmosphere) is almost constant at 1.4 kJ m⁻² s⁻¹. Most of this incoming sunlight energy is reflected by the atmosphere, heats the atmosphere and Earth's surface or causes the evaporation of water. Less than 0.1% actually falls on leaves and is thus available for photosynthesis.

The **net primary production (NPP)** is the energy available for consumption by the heterotrophes. NPP can therefore be used to compare the productivity of different ecosystems. For example:

ECOSYSTEM	NPP (arbitrary units)
Coral reef	1000
Rainforest	880
Estuaries	600
Deciduous forest	500
Grassland	260
Open ocean	50
Desert	2

Productivity may be expressed as **units of energy** (e.g. kJ m⁻² yr⁻¹) or **units of mass** (e.g. kg m⁻² yr⁻¹).

Energy transfer from producer to primary consumer is typically in the order of 5–10% of NPP.
This is because
1. Much of plant biomass (NPP) is indigestible to herbivores – there are no animal enzymes to digest lignin and cellulose.
2. Much of the plant biomass may not be consumed by any individual herbivore species – roots may be inaccessible or trampled grass may be considered uneatable.

Energy transfer from primary consumer (herbivore) to secondary consumer (carnivore) is typically 10–20% of herbivore biomass. This is more efficient than producer → consumer because
1. animal tissue is more digestible than plant tissue;
2. animal tissue has a higher energy value;
3. carnivores may be extremely specialized for prey consumption;
but is still considerably less than 100% because
a. some animal tissue – bone, hooves and hide for example – is not readily digestible;
b. feeding is not 100% efficient – much digestible material (e.g. food fragments and blood) may be lost to the environment.

PRODUCER

PRIMARY CONSUMER

SECONDARY CONSUMER

DECOMPOSER ORGANISMS

The **decomposers** are fungi and bacteria which obtain energy and raw materials from animal and plant remains. In some situations 80% or more of the productivity at any trophic level may go through a decomposer pathway (e.g. forest floors of tropical forests). In some ecosystems – peat bogs, for example – the cold, wet, acidic conditions inhibit decomposition to such an extent that only about 10% of the material entering the decomposer food chain is broken down. The remainder accumulates as peat.

The limit to the number of trophic levels is determined by:
1. the total producer biomass;
2. the efficiency of energy transfer between trophic levels (only 10%).

In practice, the energy losses limit the number of levels to 3 or 4, very rarely 5 or 6. The longest food chains can only be supported by an enormous producer biomass, e.g. a 6 level chain will only have about 10% x 10% x 10% x 10% of NPP available to the top carnivores. The enormous volume of the oceans can provide sufficient biomass to support the longest food chains.

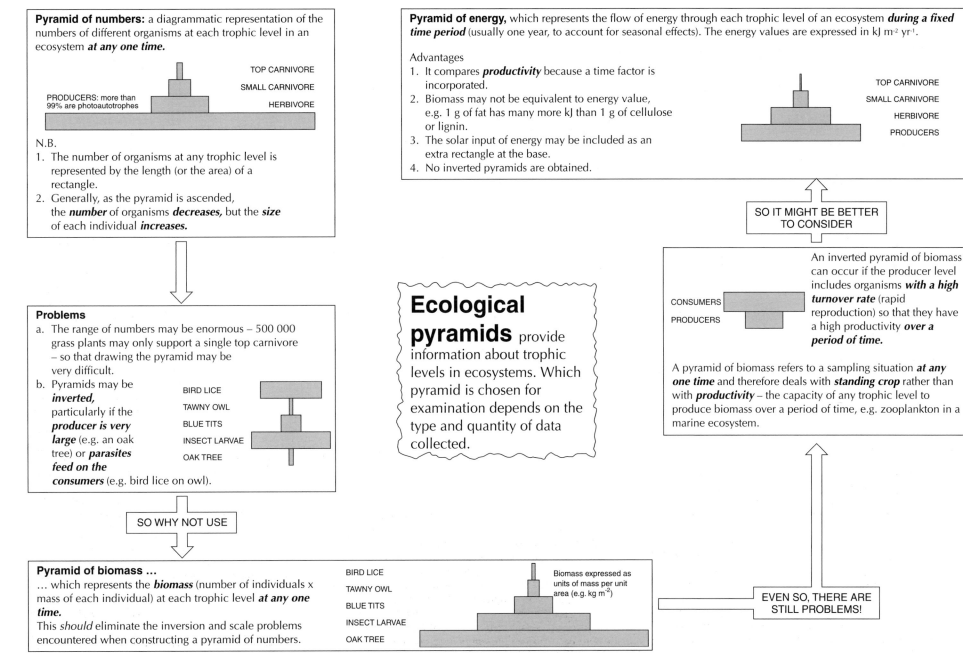

Pyramid of numbers: a diagrammatic representation of the numbers of different organisms at each trophic level in an ecosystem *at any one time.*

TOP CARNIVORE
SMALL CARNIVORE
HERBIVORE
PRODUCERS: more than 99% are photoautotrophes

N.B.
1. The number of organisms at any trophic level is represented by the length (or the area) of a rectangle.
2. Generally, as the pyramid is ascended, the *number* of organisms *decreases,* but the *size* of each individual *increases.*

Problems
a. The range of numbers may be enormous – 500 000 grass plants may only support a single top carnivore – so that drawing the pyramid may be very difficult.
b. Pyramids may be *inverted,* particularly if the *producer is very large* (e.g. an oak tree) or *parasites feed on the consumers* (e.g. bird lice on owl).

BIRD LICE
TAWNY OWL
BLUE TITS
INSECT LARVAE
OAK TREE

SO WHY NOT USE

Pyramid of biomass …
… which represents the *biomass* (number of individuals x mass of each individual) at each trophic level *at any one time.*
This *should* eliminate the inversion and scale problems encountered when constructing a pyramid of numbers.

BIRD LICE
TAWNY OWL
BLUE TITS
INSECT LARVAE
OAK TREE

Biomass expressed as units of mass per unit area (e.g. kg m^{-2})

Ecological pyramids provide information about trophic levels in ecosystems. Which pyramid is chosen for examination depends on the type and quantity of data collected.

Pyramid of energy, which represents the flow of energy through each trophic level of an ecosystem *during a fixed time period* (usually one year, to account for seasonal effects). The energy values are expressed in kJ m^{-2} yr^{-1}.

Advantages
1. It compares *productivity* because a time factor is incorporated.
2. Biomass may not be equivalent to energy value, e.g. 1 g of fat has many more kJ than 1 g of cellulose or lignin.
3. The solar input of energy may be included as an extra rectangle at the base.
4. No inverted pyramids are obtained.

TOP CARNIVORE
SMALL CARNIVORE
HERBIVORE
PRODUCERS

SO IT MIGHT BE BETTER TO CONSIDER

An inverted pyramid of biomass can occur if the producer level includes organisms *with a high turnover rate* (rapid reproduction) so that they have a high productivity *over a period of time.*

CONSUMERS
PRODUCERS

A pyramid of biomass refers to a sampling situation *at any one time* and therefore deals with *standing crop* rather than with *productivity* – the capacity of any trophic level to produce biomass over a period of time, e.g. zooplankton in a marine ecosystem.

EVEN SO, THERE ARE STILL PROBLEMS!

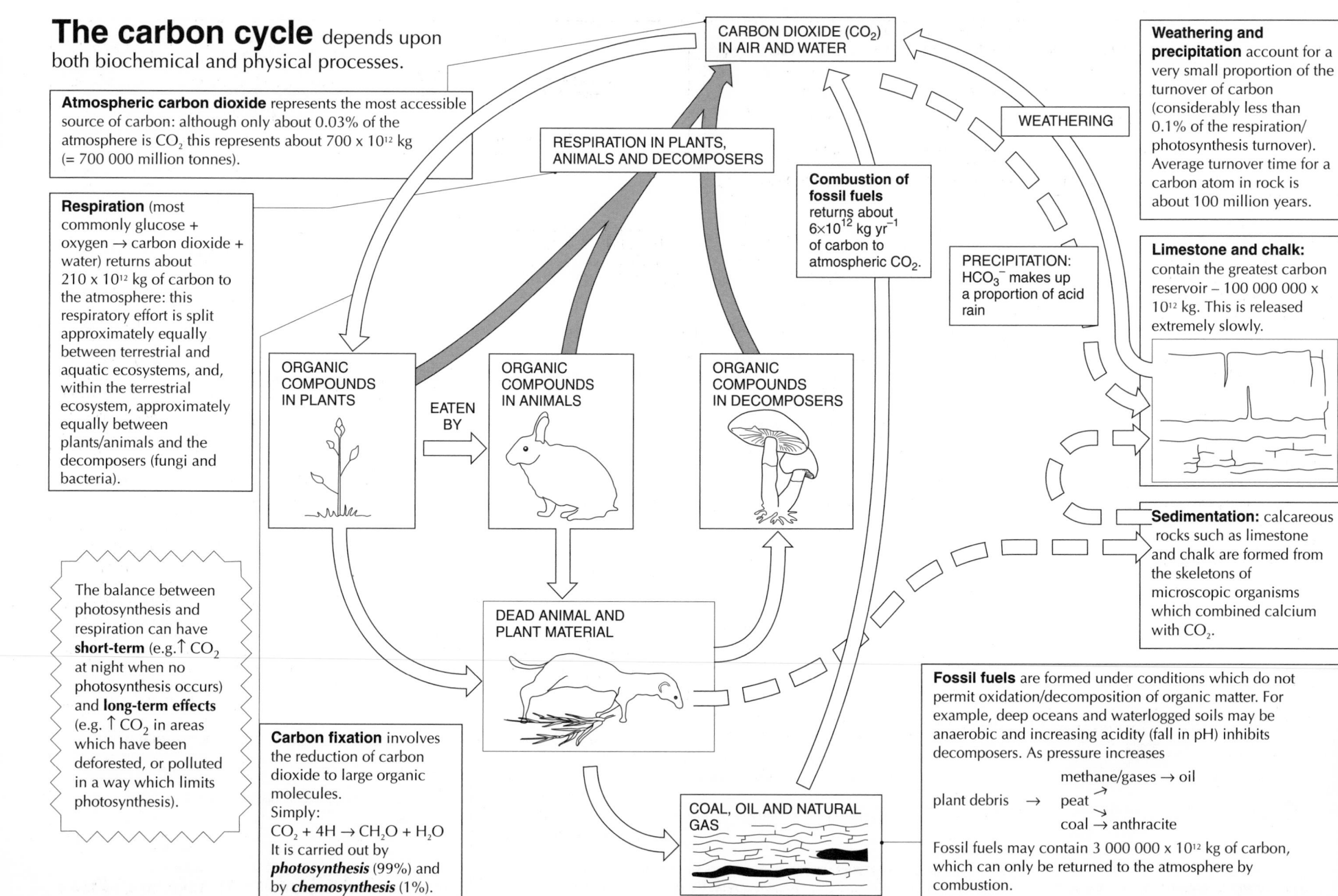

The carbon cycle depends upon both biochemical and physical processes.

Atmospheric carbon dioxide represents the most accessible source of carbon: although only about 0.03% of the atmosphere is CO_2 this represents about 700×10^{12} kg (= 700 000 million tonnes).

Respiration (most commonly glucose + oxygen → carbon dioxide + water) returns about 210×10^{12} kg of carbon to the atmosphere: this respiratory effort is split approximately equally between terrestrial and aquatic ecosystems, and, within the terrestrial ecosystem, approximately equally between plants/animals and the decomposers (fungi and bacteria).

CARBON DIOXIDE (CO_2) IN AIR AND WATER

RESPIRATION IN PLANTS, ANIMALS AND DECOMPOSERS

Combustion of fossil fuels returns about 6×10^{12} kg yr^{-1} of carbon to atmospheric CO_2.

WEATHERING

Weathering and precipitation account for a very small proportion of the turnover of carbon (considerably less than 0.1% of the respiration/ photosynthesis turnover). Average turnover time for a carbon atom in rock is about 100 million years.

PRECIPITATION: HCO_3^- makes up a proportion of acid rain

Limestone and chalk: contain the greatest carbon reservoir – 100 000 000 x 10^{12} kg. This is released extremely slowly.

ORGANIC COMPOUNDS IN PLANTS

EATEN BY

ORGANIC COMPOUNDS IN ANIMALS

ORGANIC COMPOUNDS IN DECOMPOSERS

Sedimentation: calcareous rocks such as limestone and chalk are formed from the skeletons of microscopic organisms which combined calcium with CO_2.

The balance between photosynthesis and respiration can have **short-term** (e.g.↑ CO_2 at night when no photosynthesis occurs) and **long-term effects** (e.g. ↑ CO_2 in areas which have been deforested, or polluted in a way which limits photosynthesis).

DEAD ANIMAL AND PLANT MATERIAL

Carbon fixation involves the reduction of carbon dioxide to large organic molecules. Simply: $CO_2 + 4H \rightarrow CH_2O + H_2O$ It is carried out by **photosynthesis** (99%) and by **chemosynthesis** (1%).

COAL, OIL AND NATURAL GAS

Fossil fuels are formed under conditions which do not permit oxidation/decomposition of organic matter. For example, deep oceans and waterlogged soils may be anaerobic and increasing acidity (fall in pH) inhibits decomposers. As pressure increases

plant debris → peat → methane/gases → oil, coal → anthracite

Fossil fuels may contain 3 000 000 x 10^{12} kg of carbon, which can only be returned to the atmosphere by combustion.

The nitrogen cycle depends on micro-organisms.

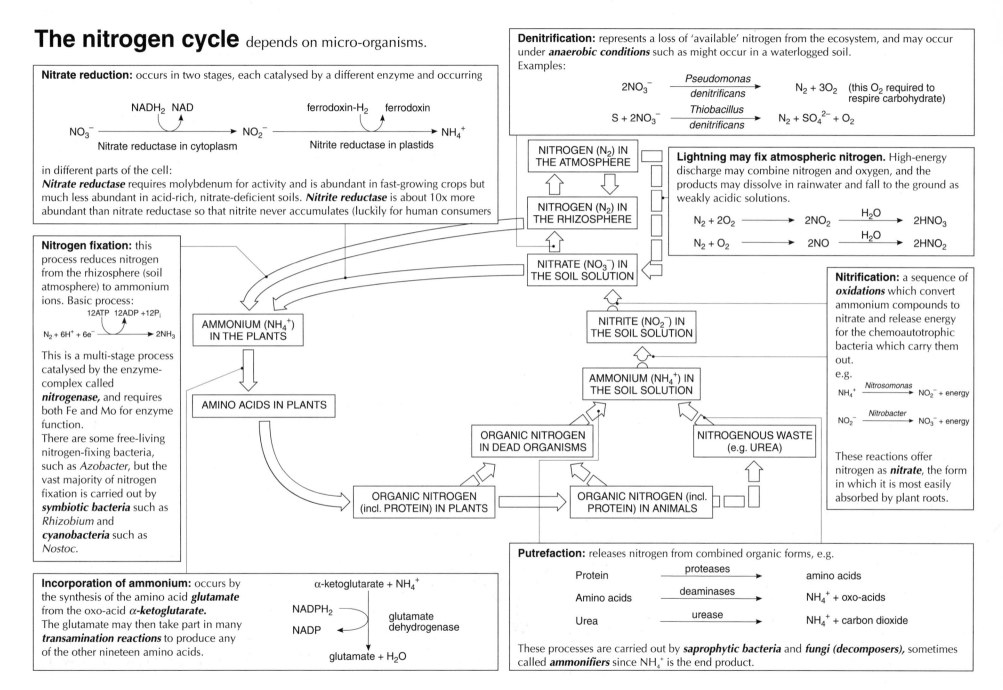

Nitrate reduction: occurs in two stages, each catalysed by a different enzyme and occurring

$$NO_3^- \xrightarrow[\text{Nitrate reductase in cytoplasm}]{NADH_2 \quad NAD} NO_2^- \xrightarrow[\text{Nitrite reductase in plastids}]{ferrodoxin\text{-}H_2 \quad ferrodoxin} NH_4^+$$

in different parts of the cell:
Nitrate reductase requires molybdenum for activity and is abundant in fast-growing crops but much less abundant in acid-rich, nitrate-deficient soils. **Nitrite reductase** is about 10x more abundant than nitrate reductase so that nitrite never accumulates (luckily for human consumers

Nitrogen fixation: this process reduces nitrogen from the rhizosphere (soil atmosphere) to ammonium ions. Basic process:

$$N_2 + 6H^+ + 6e^- \xrightarrow[]{12ATP \quad 12ADP + 12P_i} 2NH_3$$

This is a multi-stage process catalysed by the enzyme-complex called **nitrogenase,** and requires both Fe and Mo for enzyme function.
There are some free-living nitrogen-fixing bacteria, such as *Azobacter*, but the vast majority of nitrogen fixation is carried out by **symbiotic bacteria** such as *Rhizobium* and **cyanobacteria** such as *Nostoc*.

Incorporation of ammonium: occurs by the synthesis of the amino acid **glutamate** from the oxo-acid *α-ketoglutarate.*
The glutamate may then take part in many **transamination reactions** to produce any of the other nineteen amino acids.

$$\alpha\text{-ketoglutarate} + NH_4^+$$
$$NADPH_2 \searrow$$
$$NADP \xleftarrow{\quad} \text{glutamate dehydrogenase}$$
$$\text{glutamate} + H_2O$$

Denitrification: represents a loss of 'available' nitrogen from the ecosystem, and may occur under **anaerobic conditions** such as might occur in a waterlogged soil. Examples:

$$2NO_3^- \xrightarrow[\text{denitrificans}]{\text{Pseudomonas}} N_2 + 3O_2 \quad \text{(this O}_2\text{ required to respire carbohydrate)}$$

$$S + 2NO_3^- \xrightarrow[\text{denitrificans}]{\text{Thiobacillus}} N_2 + SO_4^{2-} + O_2$$

Lightning may fix atmospheric nitrogen. High-energy discharge may combine nitrogen and oxygen, and the products may dissolve in rainwater and fall to the ground as weakly acidic solutions.

$$N_2 + 2O_2 \longrightarrow 2NO_2 \xrightarrow{H_2O} 2HNO_3$$
$$N_2 + O_2 \longrightarrow 2NO \xrightarrow{H_2O} 2HNO_2$$

Nitrification: a sequence of **oxidations** which convert ammonium compounds to nitrate and release energy for the chemoautotrophic bacteria which carry them out. e.g.

$$NH_4^+ \xrightarrow{\text{Nitrosomonas}} NO_2^- + energy$$
$$NO_2^- \xrightarrow{\text{Nitrobacter}} NO_3^- + energy$$

These reactions offer nitrogen as **nitrate,** the form in which it is most easily absorbed by plant roots.

Putrefaction: releases nitrogen from combined organic forms, e.g.

Protein	$\xrightarrow{\text{proteases}}$	amino acids
Amino acids	$\xrightarrow{\text{deaminases}}$	$NH_4^+ + \text{oxo-acids}$
Urea	$\xrightarrow{\text{urease}}$	$NH_4^+ + \text{carbon dioxide}$

These processes are carried out by **saprophytic bacteria** and **fungi (decomposers),** sometimes called **ammonifiers** since NH_4^+ is the end product.

Diagram boxes:
NITROGEN (N_2) IN THE ATMOSPHERE
NITROGEN (N_2) IN THE RHIZOSPHERE
NITRATE (NO_3^-) IN THE SOIL SOLUTION
AMMONIUM (NH_4^+) IN THE PLANTS
AMINO ACIDS IN PLANTS
ORGANIC NITROGEN (incl. PROTEIN) IN PLANTS
ORGANIC NITROGEN IN DEAD ORGANISMS
NITRITE (NO_2^-) IN THE SOIL SOLUTION
AMMONIUM (NH_4^+) IN THE SOIL SOLUTION
ORGANIC NITROGEN (incl. PROTEIN) IN ANIMALS
NITROGENOUS WASTE (e.g. UREA)

The greenhouse effect is a

natural feature of the Earth, but when upset may lead to *global warming.*

ORIGINS OF GREENHOUSE GASES

Photosynthesis in forests and grasslands removes carbon dioxide (CO_2) from the atmosphere.

Car exhaust emissions contain much CO_2 – released to the atmosphere.

Combustion of fossil fuels by industrial plants releases large amounts of CO_2.

Phew!

Ruminant fermentation produces **methane** (CH_4) which cattle release into the atmosphere. Intensive cattle ranching increases CH_4 release at the expense of CO_2 uptake by photosynthesis.

Aerosol propellants contain **chlorofluorocarbons** (CFCs) which are 10^5 x worse than carbon dioxide as greenhouse gases.

Anaerobic fermentation in swamps and paddy fields produces CH_4. Inorganic fertilizers cause release of nitric oxide (NO).

All living organisms release carbon dioxide by respiration – the additional **greenhouse gases** contributed by humans *(anthropogenic contributions)* include methane and CFCs in addition to greater quantities of carbon dioxide.

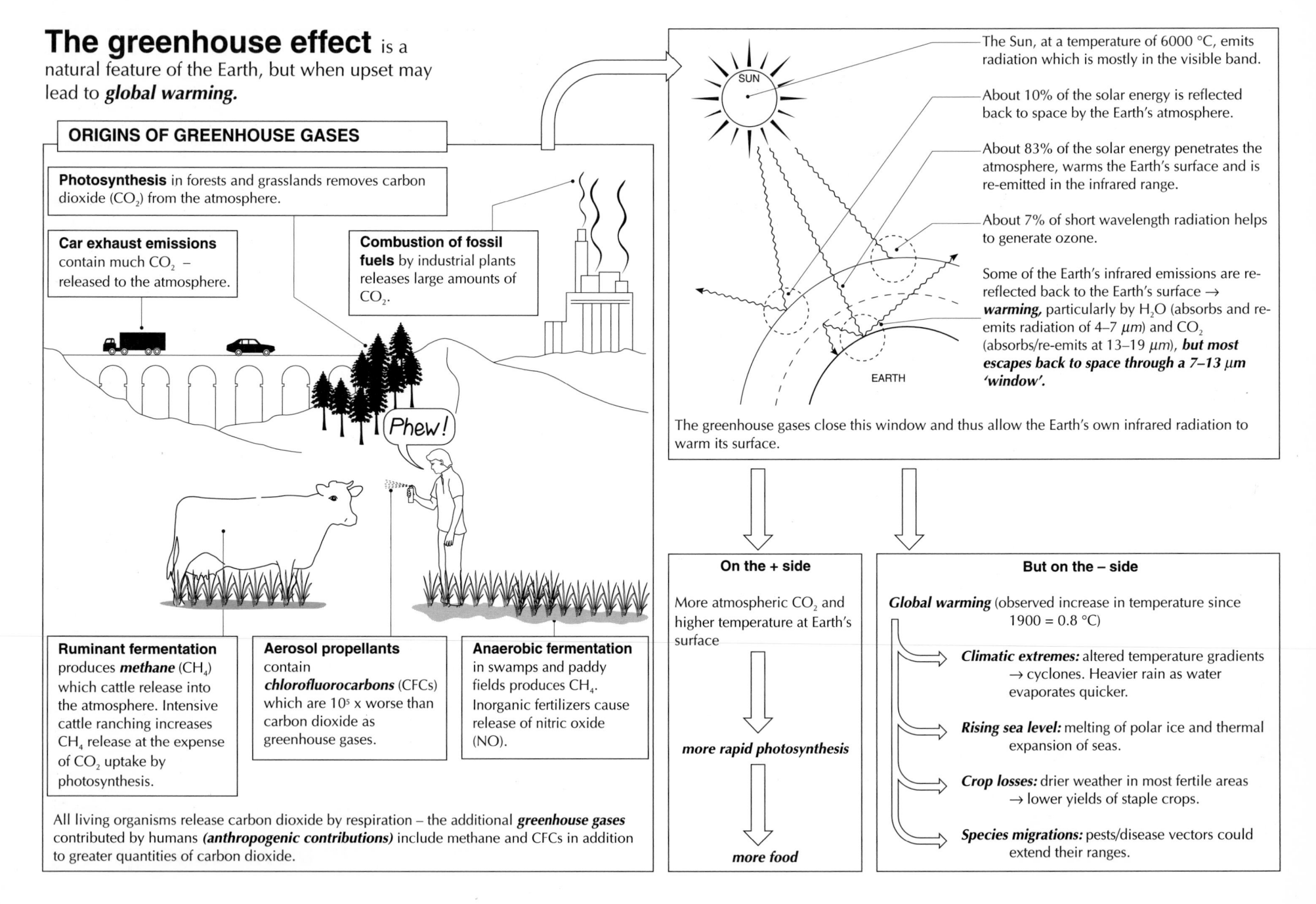

The Sun, at a temperature of 6000 °C, emits radiation which is mostly in the visible band.

About 10% of the solar energy is reflected back to space by the Earth's atmosphere.

About 83% of the solar energy penetrates the atmosphere, warms the Earth's surface and is re-emitted in the infrared range.

About 7% of short wavelength radiation helps to generate ozone.

Some of the Earth's infrared emissions are re-reflected back to the Earth's surface → *warming,* particularly by H_2O (absorbs and re-emits radiation of 4–7 μm) and CO_2 (absorbs/re-emits at 13–19 μm), *but most escapes back to space through a 7–13 μm 'window'.*

EARTH

The greenhouse gases close this window and thus allow the Earth's own infrared radiation to warm its surface.

On the + side

More atmospheric CO_2 and higher temperature at Earth's surface

more rapid photosynthesis

more food

But on the – side

Global warming (observed increase in temperature since 1900 = 0.8 °C)

Climatic extremes: altered temperature gradients → cyclones. Heavier rain as water evaporates quicker.

Rising sea level: melting of polar ice and thermal expansion of seas.

Crop losses: drier weather in most fertile areas → lower yields of staple crops.

Species migrations: pests/disease vectors could extend their ranges.

Deforestation: the rapid destruction of woodland.

BULLDOZE, SLASH AND BURN

- Has been occurring on a major scale throughout the world.

- Between 1880 and 1980 about 40% of all tropical rainforest was destroyed.

- Britain has fallen from 85% forest cover to about 8% (probably the lowest in Europe).

- Major reasons
  - -removal of hardwood for high-quality furnishings;
  - -removal of softwoods for chipboards, paper and other wood products;
  - -clearance for cattle ranching and for cash crop agriculture;
  - -clearance for urban development (roads and towns being built).

Current losses: per minute: about 11 hectares **per minute** (that's about 40 soccer or hockey pitches!).

Reduction in soil fertility

1. Deciduous trees may contain 90% of the nutrients in a forest ecosystem: these nutrients are removed, and are thus not available to the soil, if the trees are cut down and taken away.

2. Soil erosion may be rapid since in the absence of trees
   a. wind and direct rain may remove the soil;
   b. soil structure is no longer stabilized by tree root systems.

N.B. The soil below coniferous forests is often of poor quality for agriculture because the shed pine needles contain toxic compounds which act as germination and growth inhibitors.

Flooding and landslips

Heavy rainfall on deforested land is not 'held up': normally 25% of rainfall is absorbed by foliage or evaporates and 50% is absorbed by root systems. As a result water may accumulate rapidly in river valleys, often causing landslips from steep hillsides.

Changes in recycling of materials: Fewer trees mean

1. atmospheric CO_2 concentration may rise as less CO_2 is removed for photosynthesis;
2. atmospheric O_2 – vital for aerobic respiration – is diminished as less is produced by photosynthesis;
3. the atmosphere may become drier and the soil wetter as evaporation (from soil) is slower than transpiration (from trees).

Climatic changes

1. Reduced transpiration rates and drier atmosphere affect the water cycle and reduce rainfall.
2. Rapid heat absorption by bare soil raises the temperature of the lower atmosphere in some areas, causing thermal gradients which result in more frequent and intense winds.

Many plant species may have medicinal properties,
e.g. as tranquillizers, reproductive hormones, anticoagulants, painkillers and antibiotics. The Madagascan periwinkle, for example, yields one of the most potent known anti-leukaemia drugs.

Species extinction: Many species are dependent on forest conditions.

e.g. mountain gorilla depends on cloud forest of Central Africa; golden lion tamarin depends on coastal rainforest of Brazil; osprey depends on mature pine forests in Northern Europe.

It is estimated that one plant and one animal species become extinct every 30 minutes due to deforestation.

Human digestive system: I

Palate: separates breathing and feeding pathways, allowing both processes to go on simultaneously so different types of teeth evolved.

Teeth: cut, tear and grind food so that solid foods are reduced to smaller particles for swallowing, and the food has a larger surface area for enzyme action.

Salivary glands: produce *saliva,* which is 99% water plus mucin, chloride ions (activate amylase), hydrogen carbonate and phosphate (maintain pH about 6.5), lysozyme and salivary amylase.

Paratoid

Sublingual

Submandibular

Tongue: manoeuvres food for chewing and rolls food into a bolus for swallowing. Mixes food with saliva.

Diaphragm: a muscular 'sheet' separating the thorax and abdomen.

Liver: an accessory organ which produces bile and stores it in the gall bladder.

Bile duct: carries bile from gall bladder to duodenum.

Pancreas: an accessory organ producing a wide range of digestive secretions, as well as hormones.

Duodenum: the first 30 cm of the small intestine. Receives pancreatic secretions and bile and produces an alkaline mucus for protection, lubrication and chyme neutralization.

Ileum: up to 6 m in length – main site for absorption of soluble products of digestion.

Appendix: no function in humans. It is a vestige of the caecum in other mammals (herbivores).

Anal sphincter: regulates release of faeces (defecation).

Uvula: extension of soft palate which separates nasal chamber from pharynx.

Epiglottis: muscular flap which reflexly closes the trachea during swallowing to prevent food entry to respiratory tree.

Oesophagus: muscular tube which is dorsal to the trachea and connects the buccal cavity to the stomach. Muscular to generate peristaltic waves, which drive bolus of food downwards, and glandular to lubricate bolus with mucus. Semi-solid food passes to stomach in 4–8 seconds, very soft foods and liquids take only 1 second.

Cardiac sphincter: allows entry of food to stomach. Helps to retain food in stomach.

Stomach: a muscular bag which is distensible to permit storage of large quantities of food. Mucosal lining is glandular, with numerous gastric pits which secrete digestive juices. Three muscle layers including an oblique layer churn the stomach contents to ensure thorough mixing and eventual transfer of chyme to the duodenum.

Pyloric sphincter: opens to permit passage of chyme into duodenum and closes to prevent backflow of food from duodenum to stomach.

Colon (large intestine): absorbs water from faeces. Some B vitamins and vitamin K are synthesized by colonic bacteria. Mucus glands lubricate faeces.

Rectum: stores faeces before expulsion.

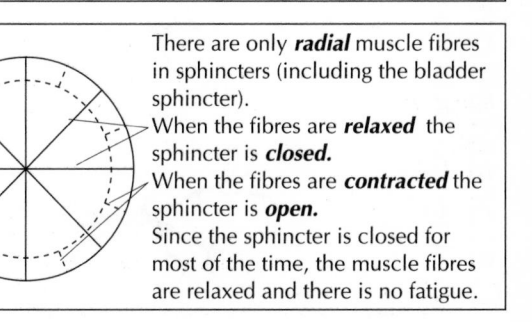

There are only *radial* muscle fibres in sphincters (including the bladder sphincter).
When the fibres are *relaxed* the sphincter is *closed.*
When the fibres are *contracted* the sphincter is *open.*
Since the sphincter is closed for most of the time, the muscle fibres are relaxed and there is no fatigue.

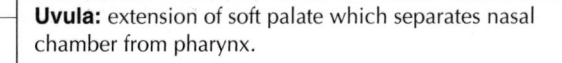

Human digestive system: II

Digestion of protein, fat and carbohydrate

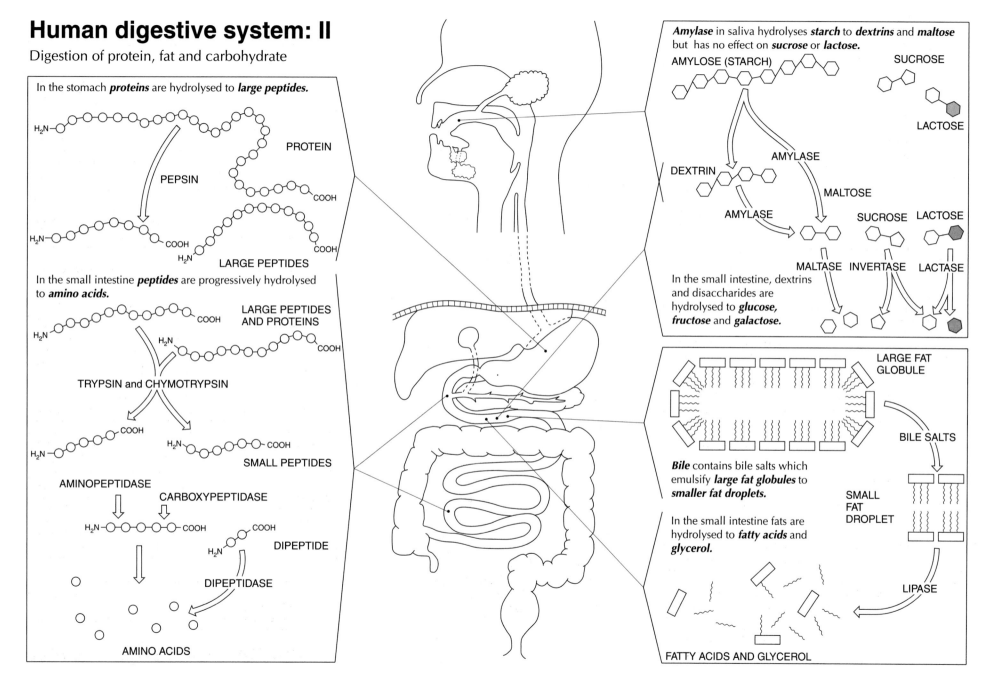

In the stomach **proteins** are hydrolysed to **large peptides**.

PROTEIN

PEPSIN

LARGE PEPTIDES

In the small intestine **peptides** are progressively hydrolysed to **amino acids**.

LARGE PEPTIDES AND PROTEINS

TRYPSIN and CHYMOTRYPSIN

SMALL PEPTIDES

AMINOPEPTIDASE

CARBOXYPEPTIDASE

DIPEPTIDE

DIPEPTIDASE

AMINO ACIDS

Amylase in saliva hydrolyses **starch** to **dextrins** and **maltose** but has no effect on **sucrose** or **lactose**.

AMYLOSE (STARCH)

SUCROSE

LACTOSE

DEXTRIN

AMYLASE

MALTOSE

AMYLASE

SUCROSE

LACTOSE

MALTASE   INVERTASE   LACTASE

In the small intestine, dextrins and disaccharides are hydrolysed to **glucose, fructose** and **galactose**.

LARGE FAT GLOBULE

BILE SALTS

Bile contains bile salts which emulsify **large fat globules** to **smaller fat droplets**.

SMALL FAT DROPLET

In the small intestine fats are hydrolysed to **fatty acids** and **glycerol**.

LIPASE

FATTY ACIDS AND GLYCEROL

Transverse section of generalised gut

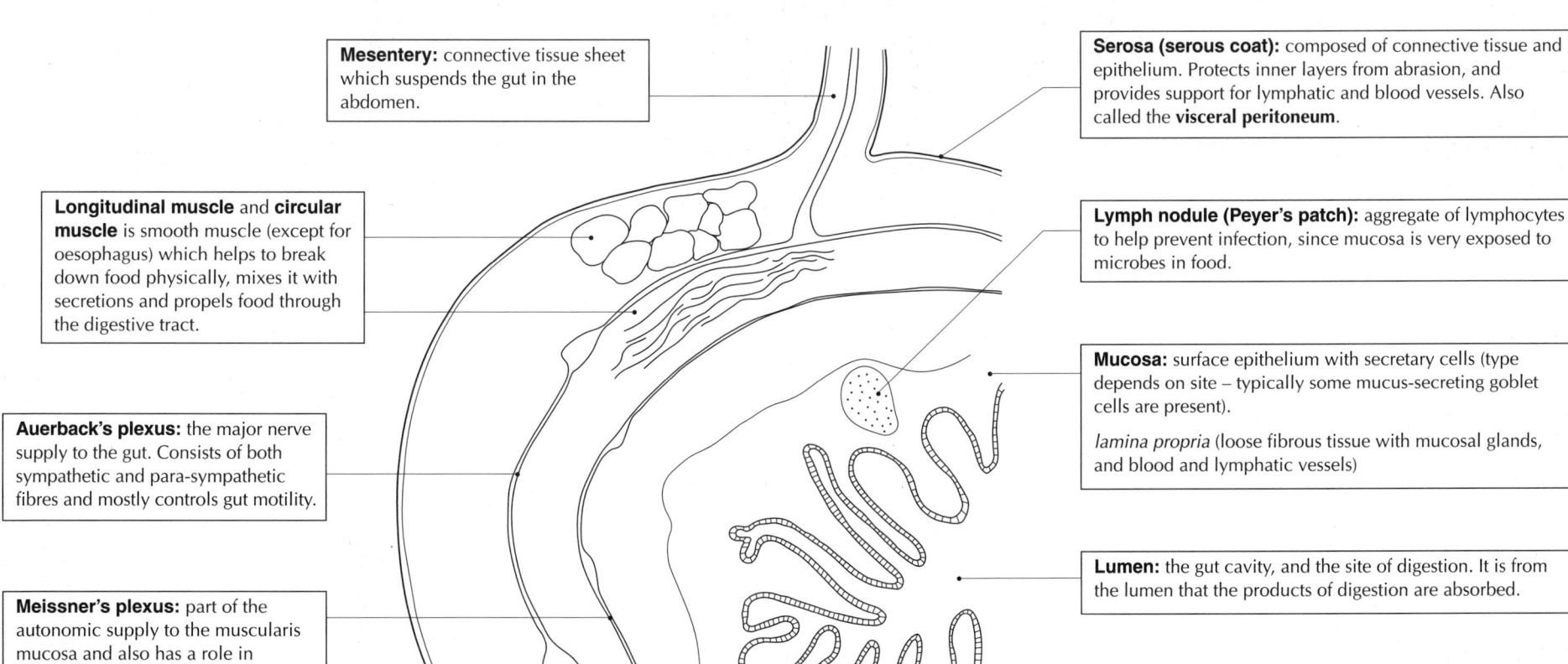

Mesentery: connective tissue sheet which suspends the gut in the abdomen.

Longitudinal muscle and **circular muscle** is smooth muscle (except for oesophagus) which helps to break down food physically, mixes it with secretions and propels food through the digestive tract.

Auerback's plexus: the major nerve supply to the gut. Consists of both sympathetic and para-sympathetic fibres and mostly controls gut motility.

Meissner's plexus: part of the autonomic supply to the muscularis mucosa and also has a role in controlling secretions by the gut.

External (accessory) gland: salivary glands, pancreas and liver all produce secretions which are poured into the lumen via ducts.

Serosa (serous coat): composed of connective tissue and epithelium. Protects inner layers from abrasion, and provides support for lymphatic and blood vessels. Also called the **visceral peritoneum**.

Lymph nodule (Peyer's patch): aggregate of lymphocytes to help prevent infection, since mucosa is very exposed to microbes in food.

Mucosa: surface epithelium with secretary cells (type depends on site – typically some mucus-secreting goblet cells are present).

lamina propria (loose fibrous tissue with mucosal glands, and blood and lymphatic vessels)

Lumen: the gut cavity, and the site of digestion. It is from the lumen that the products of digestion are absorbed.

Muscularis mucosa: a thin muscle layer which can help local movements of areas of villi.

Submucosa: contains dense connective tissue and some adipose tissue which supports the nervous and blood supply for the mucosa.

Submucosal gland: only in the duodenum, when the **Brunner's glands** produce an alkaline, mucus secretion.

Absorption of the products of digestion

is aided by a large surface area, specific uptake systems and well-developed transport network.

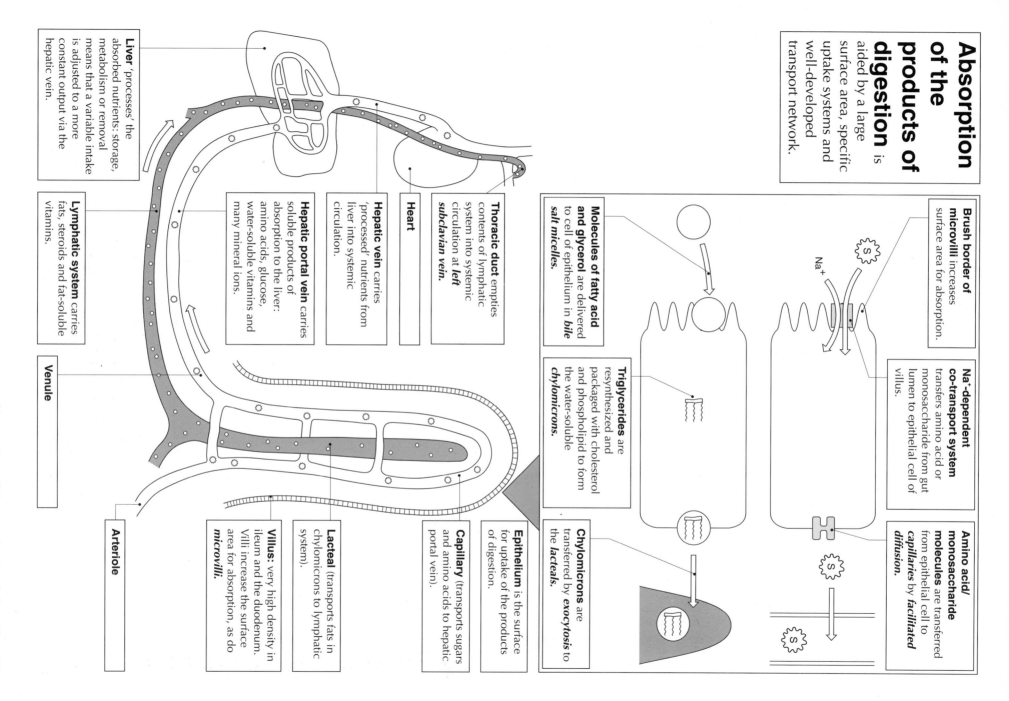

Liver 'processes' the absorbed nutrients: storage, metabolism or removal means that a variable intake is adjusted to a more constant output via the hepatic vein.

Lymphatic system carries fats, steroids and fat-soluble vitamins.

Venule

Arteriole

Hepatic portal vein carries soluble products of absorption to the liver: amino acids, glucose, water-soluble vitamins and many mineral ions.

Thoracic duct empties contents of lymphatic system into systemic circulation at **left subclavian vein.**

Hepatic vein carries 'processed' nutrients from liver into systemic circulation.

Heart

Molecules of fatty acid and glycerol are delivered to cell of epithelium in **bile salt micelles.**

Triglycerides are resynthesized and packaged with cholesterol and phospholipid to form the water-soluble **chylomicrons.**

Chylomicrons are transferred by **exocytosis** to the **lacteals.**

Epithelium is the surface for uptake of the products of digestion.

Capillary (transports sugars and amino acids to hepatic portal vein).

Lacteal (transports fats in chylomicrons to lymphatic system).

Villus: very high density in ileum and the duodenum. Villi increase the surface area for absorption, as do **microvilli.**

Brush border of microvilli increases surface area for absorption.

Na$^+$

Na$^+$-dependent co-transport system transfers amino acid or monosaccharide from gut lumen to epithelial cell of villus.

Amino acid/ monosaccharide molecules are transferred from epithelial cell to **capillaries** by **facilitated diffusion.**

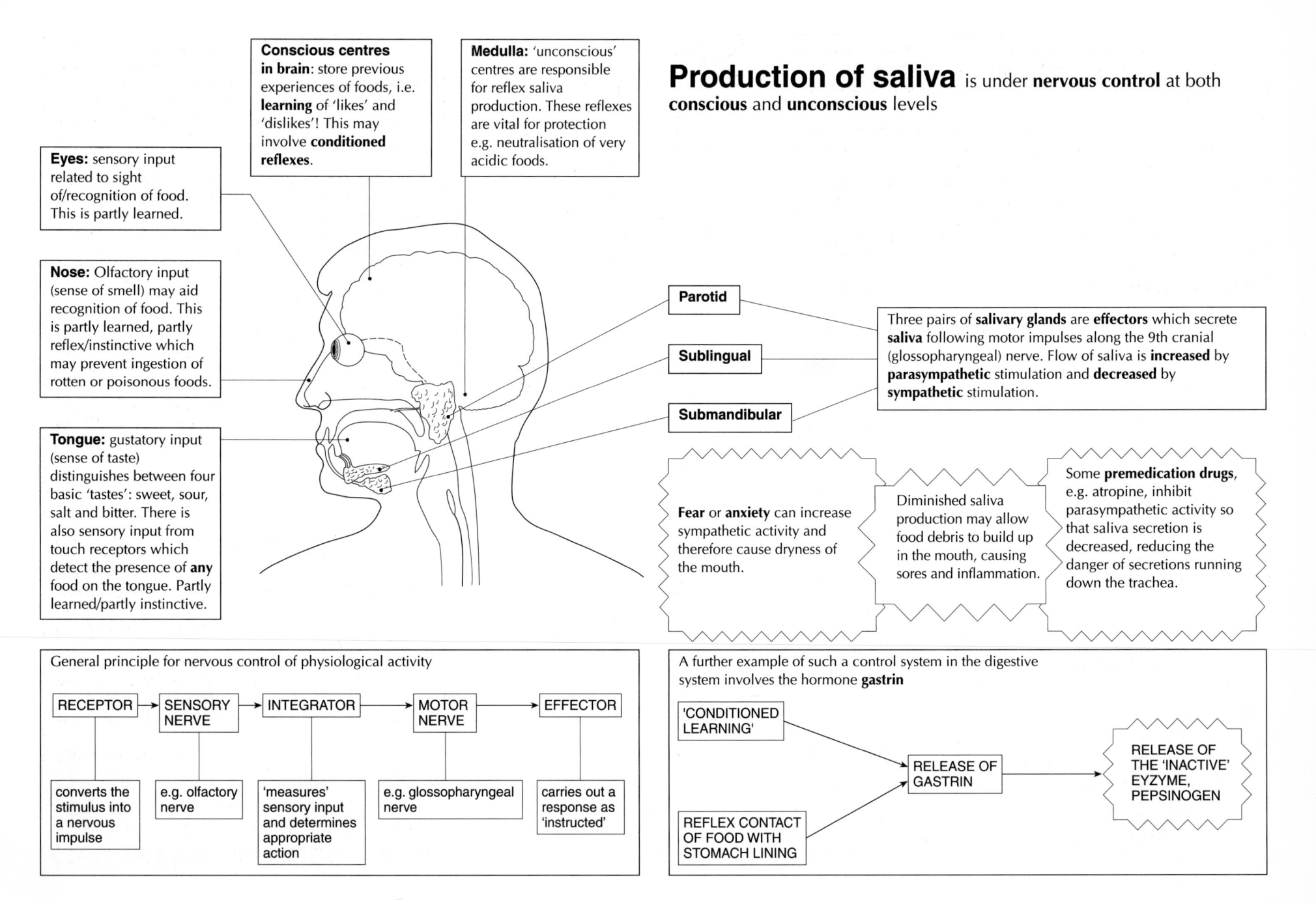

Production of saliva is under nervous control at both conscious and unconscious levels

Eyes: sensory input related to sight of/recognition of food. This is partly learned.

Nose: Olfactory input (sense of smell) may aid recognition of food. This is partly learned, partly reflex/instinctive which may prevent ingestion of rotten or poisonous foods.

Tongue: gustatory input (sense of taste) distinguishes between four basic 'tastes': sweet, sour, salt and bitter. There is also sensory input from touch receptors which detect the presence of **any** food on the tongue. Partly learned/partly instinctive.

Conscious centres in brain: store previous experiences of foods, i.e. **learning** of 'likes' and 'dislikes'! This may involve **conditioned reflexes**.

Medulla: 'unconscious' centres are responsible for reflex saliva production. These reflexes are vital for protection e.g. neutralisation of very acidic foods.

Parotid

Sublingual

Submandibular

Three pairs of **salivary glands** are **effectors** which secrete **saliva** following motor impulses along the 9th cranial (glossopharyngeal) nerve. Flow of saliva is **increased** by **parasympathetic** stimulation and **decreased** by **sympathetic** stimulation.

Fear or **anxiety** can increase sympathetic activity and therefore cause dryness of the mouth.

Diminished saliva production may allow food debris to build up in the mouth, causing sores and inflammation.

Some **premedication drugs**, e.g. atropine, inhibit parasympathetic activity so that saliva secretion is decreased, reducing the danger of secretions running down the trachea.

General principle for nervous control of physiological activity

RECEPTOR → SENSORY NERVE → INTEGRATOR → MOTOR NERVE → EFFECTOR

converts the stimulus into a nervous impulse

e.g. olfactory nerve

'measures' sensory input and determines appropriate action

e.g. glossopharyngeal nerve

carries out a response as 'instructed'

A further example of such a control system in the digestive system involves the hormone **gastrin**

'CONDITIONED LEARNING'

REFLEX CONTACT OF FOOD WITH STOMACH LINING

RELEASE OF GASTRIN → RELEASE OF THE 'INACTIVE' EYZYME, PEPSINOGEN

An ideal human diet

An ideal human diet contains fat, protein, carbohydrate, vitamins, minerals, water and fibre *in the correct proportions.*

An adequate diet provides sufficient *energy* for the performance of metabolic work, although the 'energy food' is in unspecified form.

A balanced diet provides all dietary requirements *in the correct proportions.* Ideally this would be $1/7$ *fat,* $1/7$ *protein* and $5/7$ *carbohydrate.*

Carbohydrates

Principally as a **respiratory substrate,** i.e. to be oxidized to release **energy** for active transport, synthesis of macromolecules, cell division and muscle contraction.

Common sources: rice, potatoes, wheat and other cereal grains, i.e. as **starch** and as refined sugar, **sucrose** in food sweetenings and preservatives.

Digested in duodenum and ileum and absorbed as **glucose.**

FLOUR

Lipids

Highly reduced and therefore can be oxidized to release **energy.** Also important in **cell membranes** and as a component of **steroid hormones.**

BUTTER

Common sources: meat and animal foods are rich in **saturated fats** and **cholesterol,** plant sources such as sunflower and soya are rich in **unsaturated fats.**

Digested in duodenum and ileum and absorbed as **fatty acids and glycerol.**

Vitamins have no common structure or function but are essential in small amounts to use other dietary components efficiently. **Fat-soluble vitamins** (e.g. A, D and E) are ingested with fatty foods and **water-soluble vitamins** (B group, C) are common in fruits and vegetables.

Fibre (originally known as **roughage**) is mainly cellulose from plant cell walls and is common in fresh vegetables and cereals. It **may** provide some energy but mainly serves to aid faeces formation, prevent constipation and ensure the continued health of the muscles of large intestine.

Proteins are **building blocks** for growth and repair of many body tissues (e.g. myosin in muscle, collagen in connective tissues), as **enzymes,** as **transport systems** (e.g. haemoglobin), as **hormones** (e.g. insulin) and as **antibodies.**

Common source: meat, fish, eggs and legumes/pulses. Must contain eight **essential amino acids** since humans are not able to synthesize them. Animal sources generally contain more of the essential amino acids.

BAKED BEANS

Protein quality can be assessed quantitively as

$$\text{Apparent digestibility of protein} = \frac{\text{N (nitrogen) intake} - \text{N in faeces}}{\text{N intake}}$$

and

$$\text{Biological value of growth and maintenance} = \frac{\text{N intake} - \text{N in faeces} - \text{N in urine}}{\text{N intake} - \text{N in faeces}}$$

Digested in stomach, duodenum and ileum and absorbed as **amino acids.**

Minerals have a range of **specific** roles (direct structural components, e.g. Ca^{2+} in bones; constituents of macromolecules, e.g. PO_4^{3-} in DNA; part of pumping systems, e.g. Na^+ in glucose uptake; enzyme cofactors, e.g. Fe^{3+} in catalase; electron transfer, e.g. Cu^{2+} in cytochromes) and *collectively* help to maintain solute concentrations essential for control of water movement. They are usually ingested with other foods – dairy products and meats are particularly important sources.

Water is required as a solvent, a transport medium, a substrate in hydrolytic reactions and for lubrication. A human requires 2–3 dm^2 of water daily, most commonly from drinks and liquid foods.

Dietary vitamin D is only required in the absence of sunlight

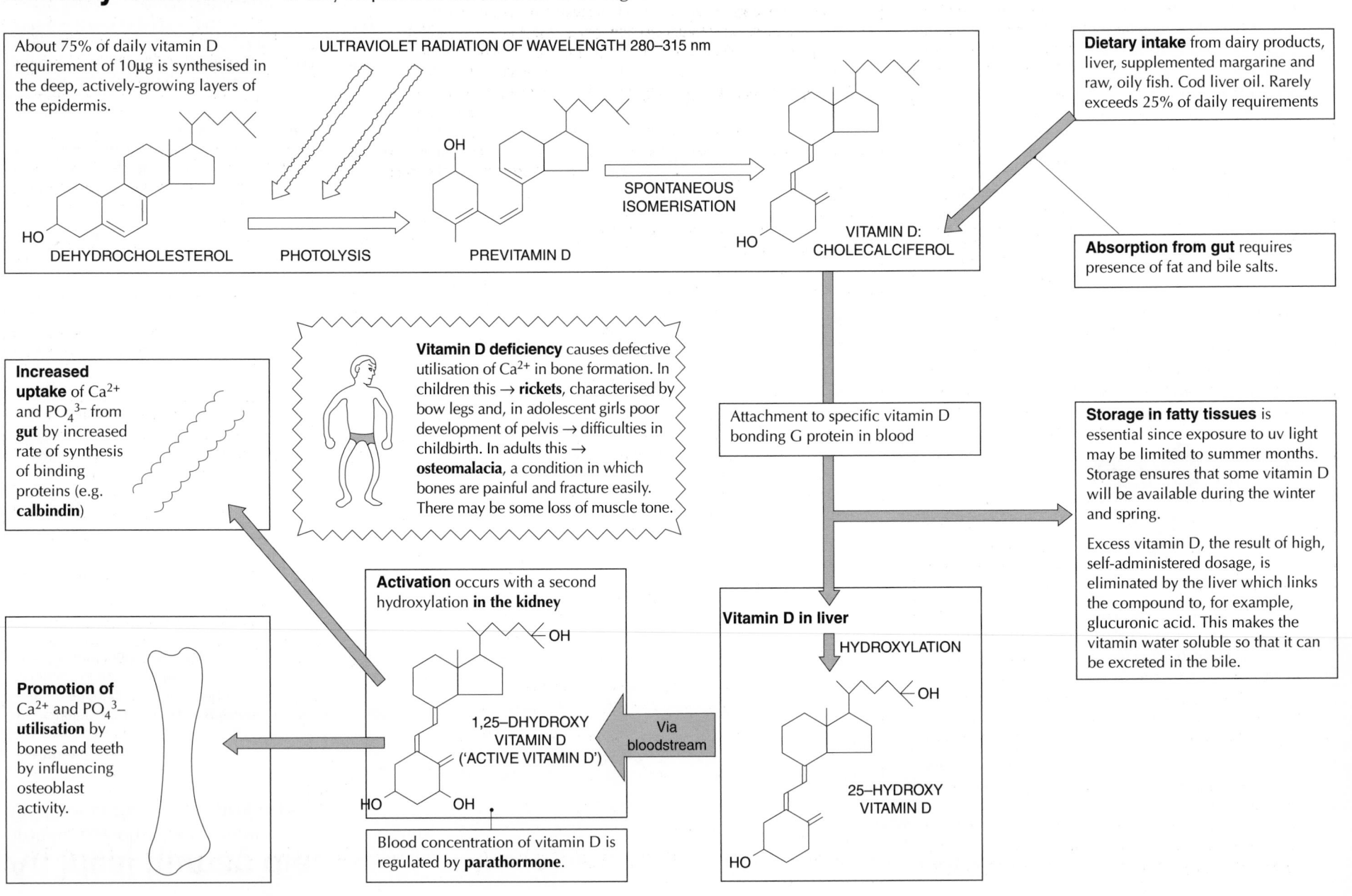

About 75% of daily vitamin D requirement of 10μg is synthesised in the deep, actively-growing layers of the epidermis.

ULTRAVIOLET RADIATION OF WAVELENGTH 280–315 nm

DEHYDROCHOLESTEROL

PHOTOLYSIS

PREVITAMIN D

SPONTANEOUS ISOMERISATION

VITAMIN D: CHOLECALCIFEROL

Dietary intake from dairy products, liver, supplemented margarine and raw, oily fish. Cod liver oil. Rarely exceeds 25% of daily requirements

Absorption from gut requires presence of fat and bile salts.

Increased uptake of Ca^{2+} and PO_4^{3-} from **gut** by increased rate of synthesis of binding proteins (e.g. **calbindin**)

Vitamin D deficiency causes defective utilisation of Ca^{2+} in bone formation. In children this → **rickets**, characterised by bow legs and, in adolescent girls poor development of pelvis → difficulties in childbirth. In adults this → **osteomalacia**, a condition in which bones are painful and fracture easily. There may be some loss of muscle tone.

Attachment to specific vitamin D bonding G protein in blood

Storage in fatty tissues is essential since exposure to uv light may be limited to summer months. Storage ensures that some vitamin D will be available during the winter and spring.

Excess vitamin D, the result of high, self-administered dosage, is eliminated by the liver which links the compound to, for example, glucuronic acid. This makes the vitamin water soluble so that it can be excreted in the bile.

Activation occurs with a second hydroxylation **in the kidney**

1,25–DHYDROXY VITAMIN D ('ACTIVE VITAMIN D')

Via bloodstream

Vitamin D in liver

HYDROXYLATION

25–HYDROXY VITAMIN D

Promotion of Ca^{2+} and PO_4^{3-} **utilisation** by bones and teeth by influencing osteoblast activity.

Blood concentration of vitamin D is regulated by **parathormone**.

BENEFITS OF THE VEGETARIAN DIET

Current research suggests a reduction in

- hypertension
- colon cancer
- breast cancer
- mature-onset diabetes
- diverticuiitis
- dental caries
- cardiovascular disease

In vegetarians, especially those who neither drink nor smoke. These benefits are thought to be largely due to

- increased content of dietary fibre
- higher ratio of unsaturated: saturated fatty acids

Vegetarians may choose a diet of grains, fruits, vegetables and nuts for ethical or nutritional reasons.

A vegetarian diet must, of course, satisfy the basic nutritional demands of an animal, i.e. the provision of fat, protein, carbohydrate, vitamins, minerals, fibre and water.

This can be relatively straightforward if the diet contains a small quantity of animal products, thus

Ovolactovegetarians (who eat milk products and eggs) and
Pescovegetarians (who eat fish) have few problems maintaining a balanced diet.

DEFICIENCIES OF THE VEGETARIAN DIET

Iron: is very much more common in meats, especially liver, and anaemia is three times more common in vegetarians than in non-vegetarians. Green leafy vegetables are a good source, but menstruating women may need an iron supplement.

Calcium: often very low in vegetarian diets, especially if the hull of grains and legumes are removed. Calcium can easily be provided if milk products are part of the diet, but the problem can be made worse by the high fibre content of vegetarian diets since phytate in fibre lowers the bioavailability of calcium.

Vitamin B$_{12}$ is made by micro-organisms and, apart from wheatgerm and beer, is not present in vegetable products. Deficiency leads to pernicious anaemia, since this vitamin is essential for the production of erythrocytes.

Vitamin D: not found in plant foods, but very common in dairy products so not a problem for ovolactovegetarians. Vitamin D is synthesised in the skin if exposure to sunlight is adequate.

Protein: the most quoted failing of vegetarian diets. Vegetable protein does not contain all amino acids in the correct relative proportions for the synthesis of animal proteins – for example corn is deficient in lysine and beans in methionine. The key is to select a combination of plant foods which complement one another.

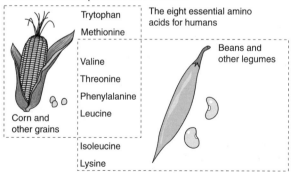

	The eight essential amino acids for humans
Trytophan	
Methionine	
Valine	Beans and other legumes
Threonine	
Phenylalanine	
Leucine	
Isoleucine	
Lysine	

Corn and other grains

Third World countries may have a vegetarian diet dictated not by choice but by sheer economics. Since the diet is likely to be dominated by a single 'energy' food, protein deficiency diseases such as Kwashiokor and Marasmus are far more likely to occur in these areas.

Fruits may be rich in sugars (glucose, fructose and sucrose), pectins and gums (soluble dietary fibre), vitamin C (particularly citrus fruits, blackcurrants, tomatoes and peppers), and provide small quantities of vitamin K.

Leaves provide dietary fibre, vitamins C, A, E and K. Some (spinach, watercress) are rich in iron.

Legumes provide protein in large amounts, carbonhydrate (energy), fibre, iron, calcium and vitamin B complex. Any vegetarian diet should contain legume material such as peas or beans.

Roots such as carrot, beet and cassava are rich in complex polysaccharides (energy),and provide small quantities of protein and vitamin C. Potato is not a root (it is a stem tuber) but as it grows underground is often treated as one.

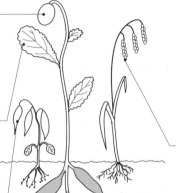

Cereals (grains) such as rice and wheat, contain carbohydrate (energy), protein, dietary fibre, vitamin B complex, iron and calcium. The vitamins and minerals are largely confined to the husk, and poor cooking or preparative techniques may considerably reduce cereal food value.

Oilseeds (sunflower, for example) provide energy (both carbohydrate and fat), some protein and a range of essential fatty acids. There is a high proportion of unsaturated fatty acids in these oils. The fat-soluble vitamins A and E may be present.

Those at greater risk on a vegetarian diet:

- **children**, since the abundance of fibrous, bulky foods may suppress appetite (that "full up" feeling) and thus limit 'energy' and protein intake.
- **expectant mothers**, with a greater dietary demand for protein, 'energy' and calcium.
- **vegans** (who eat neither fish, dairy products nor eggs) may need to supplement their diet with calcium, vitamin B$_{12}$ and vitamin D. There may also be deficiencies in some essential fatty acids which are found almost exclusively in fish.

Diets and dieting are usually associated with attempts to control body mass, although there may be other reasons for their use

'CRASH' DIETS (those which attempt to lose 2 - 3 kg per week) include

The 'Milk' Diet: comprises 1800 cm^3 (3 pints) of milk with supplementary iron, vitamins and inert bulk laxative such as bran.It provides 4900 kJ and 59g protein per day.

It is cheap and uncomplicated but very monotonous.

The VLCD (very low calorie diet): also known as the 'Cambridge' or 'protein sparing modified diet'. Provides only 1200–1800 kJ and about 40–50g of protein per day.

Sometimes recommended for the very obese, since weight loss can be rapid, but there is a loss of protein from the blood and skin, and the mass of the heart is reduced.

Considered unsuitable for infants, children, pregnant or breast-feeding women, or people with porphyria or gout. Should not be used for more than 3 or 4 weeks as the sole source of nourishment.

'Crash' diets are seldom successful in the long term. Great motivation is necessary to continue with them, they make social contacts difficult, and the body can readjust metabolic rate so that even a reduced kJ intake can lead to fat storage.

stabilisation – reduced metabolism demands fewer kJ

initial weight loss as metabolic demand for kJ exeeeds intake

further reduction in metabolic rate as body stores available kJ rather than using them

weight/kg

desired weight

10000   8000

Energy intake / kJ per day

DIETS FOR WEIGHT CONTROL

The most successful programme for weight control is one which includes

controlled diet   and   exercise

limit the intake of kjoules

increase the consumption of kjoules

The reasons for employing a weight-reducing diet may be

cosmetic: a desire to fit into a socially-acceptable body shape

health: problems associated with obesity include increased incidence of diabetes, cardiovascular accidents and arthritis

Weight reduction programmes which include an exercise regime are more likely to be successful, since

a. exercise may promote metabolism, thus increasing the kJ demand rather than allowing an adaptation of the type described opposite.

b. exercise may itself improve body form (muscle tone and posture) so that the dieter is encouraged to continue.

Statistics suggest that a dieter who is able to record a weight loss of 4–6 kg over a four week period is likely to continue with such a weight reduction programme.

MICRODIETS MAY SERVE SPECIAL FUNCTIONS

Some diets are designed not for weight loss but for reduction of some other medical problem. For example,

Gluten-free diets (gluten is a protein found in cereals, including wheat) are prescribed for sufferers which coeliac disease in which an immune response destroys the villi and causes starvation through the inability to absorb nutrients.

Reduced-sodium diets for sufferers from hypertension

Reduced-sugar diets for diabetics

Iron-enriched diets for sufferers of anaemia, particularly women at menstruation

Evening primrose oil for individuals with multiple sclerosis

Extra carrots for nocturnal bunnies!

REMEMBER – An Adequate Diet must provide

fats, **proteins**, **carbohydrates** in the correct proportions, and **vitamins**, **minerals**, **water** and **fibre**.

Typical weight reduction diets might

- limit alcohol intake

- reduce fat intake, which includes the elimination of 'hidden fat' foods such as cakes and biscuits

- reduce the intake of refined, simple sugars

- increase the intake of dietary fibre (the F-plan diet) so that food is passed through the alimentary canal at a greater rate to limit the absorption of 'potential energy'

Metabolism and metabolic rate

BASAL METABOLIC RATE (B.M.R.) is the energy consumed by the body **at rest**.

This energy is required to perform vital functions such as breathing movements, heartbeat, maintenance of ion gradients across membranes and the synthesis of molecules such as proteins

B.M.R. is measured lying down (no activity) in a warm (no thermal stress) quiet (minimal mental stress) room 12–18 hours after the last meal (no digestive activity).

is affected by a number of factors

During pregnancy and lactation there are great energy demands for production of foetal tissues and for milk.

Others which *increase* B.M.R include **food intake**, **adrenaline secretion**, **altered environmental temperature**, **exercise** and **stress**.

Other factors which *decrease* B.M.R include **fasting**, **malnutrition**, and **sleep**.

Males usually have a greater B.M.R than females since males have a greater proportion of lean body mass (fat cells have a very low metabolic rate).

As age increases B.M.R. decreases for two reasons
1 the proportion of lean body mass decreases
2 the synthesis of molecules, especially proteins, for growth decreases.

+ **Energy demands for activity** – greater as body mass increases

+ **Dietary induced thermogenesis** – digestion and absorption can demand 10% of total energy requirement

METABOLIC RATE – the energy consumed by the body during the performance of its normal activities.

B.M.R. is measured by calorimetry

Direct calorimetry

WATER

The heat released by the resting individual causes a rise in temperature e.g. of circulating water

Increased temperature of water is proportional to B.M.R.

Indirect calorimetry

A respirometer is used to measure oxygen consumption 1 dm^3 ≡ 18 kJ of energy released

B.M.R. = vol. O$_2$ consumed × 18 (per hour per m^2 body surface area)

A **Nomogram** can be used to calculate **body surface area**: a ruler should be extended from the height to weight scales – the intersect with scale III gives the body surface area e.g. the author has a surface area of 1.8 m^2.

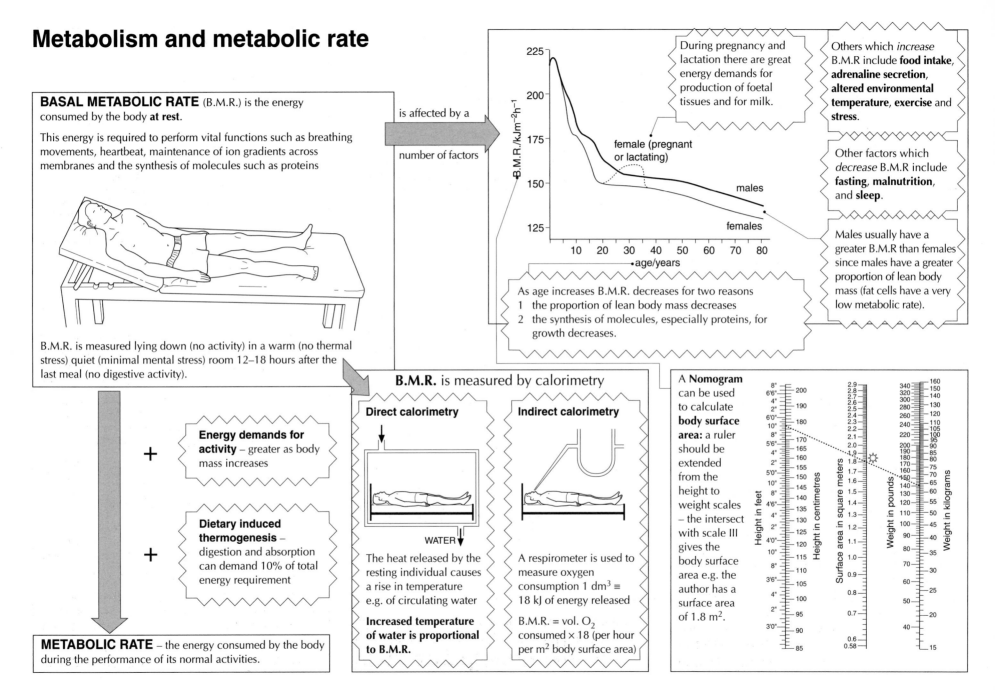

Diet and development:
the right nutrients at the right time

Pregnancy: 10% increase in both energy and protein demand, 100% increase in demand for folic acid, calcium and ascorbic acid. There is a danger if the pregnant female uses high doses of nutritional supplements – vitamins A and D can be toxic to the foetus, and high levels of some water-soluble vitamins may lead to an elevated demand in the infant after birth.

Lactation: additional nutritional demands are higher than during pregnancy – RDA's for protein and iron are increased by 20% for B vitamins and 'energy' by about 25% for ascorbic acid and vitamin A by 100% and for calcium by 150%. In addition, vitamin D supplementation is recommended.

Adolescence and puberty: both energy and protein demands may be 80% higher than in adulthood. There may be very specific demands – for example, the onset of menstruation causes pubescent females to increase demands for iron, and the thickening of bones in young males may increase the requirement for calcium. Completion of immune system may require additional ascorbic acid.

Adulthood: see earlier for general food requirements in adults. The most obvious variations are in energy demand, where activity, occupation and body mass may cause differences as great as 100% in the demand for kJoules.

The elderly: demand for energy (and associated B vitamins) and protein may fall by 25–30% or even more if a significantly more sedentary lifestyle is adopted. Some supplementation of ascorbic acid (due to a reduced intake of fresh fruit and vegetables) is recommended, and some of the symptoms of osteoporosis may be prevented by supplements of calcium and vitamin D.

Infants and children: human milk is a total food for a baby up to about 6 months of age – in the first year of life energy demands can be 200% greater than in adult life (with a proportionately higher demand for B vitamins), protein demand can be three times as great as an adult, and vitamin C, calcium, iron and vitamins A and D may be required at four times the adult rate.

Graph: Mass/kg (y-axis, marked 40 and 80) vs Time/yrs (x-axis, marked Birth, 10, 20)

Breast is best: cows milk and human milk compared

Colostrum: 'the first milk' i.e. the first secretion of the mammary glands. Produced for about five days – little energy content (and the infant may initially lose weight) but contains about 50% of the immune factors (principally immunoglobulin A) which are passed from the mother to the infant.

Human milk is low in bacteria and contains antimicrobial factors so that breast-fed infants suffer fewer infections, particularly the dangerously-dehydrating gastroenteritis. Bottle-fed infants in the Third World may have a 10–15 times greater mortality than breast-fed babies

Human milk is low-cost, is delivered at body temperature, requires no preparation and breast-feeding encourages a social bond between mother and infant. Suckling may have a contraceptive effect, although this is unclear.

Artificial milk ('formula') is based on cow's milk, but it must be modified since the gastrointestinal tract and kidneys of the human infant are immature and incapable of dealing with the "richness" of cow's milk.

Cow's milk compared with human milk

Fat	High in long-chain, saturated fatty acids	Difficult to digest inhibits calcium absorption
Protein	Three times higher	Infant kidneys cannot cope, excess amino acids may cause brain damage
	High casein: whey ratio	Causes hard curd in stomach – not easily digested
Minerals	Very high (particularly sodium)	Infant kidney cannot cope – severe (often fatal) dehydration
Lactose	Very high	Lactose intolerance – may lead to galactosaemia, the symptoms of which may include brain damage

Lung structure and function may be affected by a variety of disease conditions.

Larynx containing vocal cords which snap shut during hiccoughs.

Trachea with C-shaped rings of cartilage to support trachea in open position when thoracic pressure falls.

Lining of ciliated epithelium to trap and remove pathogens and particles of dust and smoke.

External intercostal muscles which contract to lift rib cage upwards and outwards.

Pleural membranes are moist to reduce friction as lungs move in chest.

Heart is close to lungs to drive pulmonary circulation.

Cut end of rib: ribs protect lungs and heart.

Pleural cavity at negative pressure so that passive lungs follow movement of rib cage.

Chronic bronchitis is a progressive inflammatory disease caused by exposure to irritants, including tobacco smoke, sulphur dioxide and urban fog. The mucous membrane is damaged causing swelling and fluid secretion, and reduced ciliary activity allows excess mucus to collect. There may be difficulty in breathing, and bacteria may infect the stagnant mucus, leading to pus formation. Continued shortage of oxygen leads to pulmonary hypertension and death.

Laryngitis is an inflammation of the larynx caused by a viral infection, often followed by a secondary bacterial infection.

Pleurisy is infection of the pleural membranes, causing painful breathing and impairment of the negative pressure breathing system.

Turbinate bones direct flow of air so that inspired air is warmed and moistened.

Palate separates nasal and buccal cavities to permit breathing and feeding at the same time.

Nasal hairs trap dust particles and some airborne pathogens.

Sternum for ventral attachment of ribs.

Pulmonary artery, which delivers deoxygenated blood to the lungs.

Bronchus with cartilaginous rings to prevent collapse during inspiration.

Pulmonary vein, which returns oxygenated blood to the heart.

Terminal bronchiole with no cartilage support.

Alveolus, which is the actual site of gas exchange between air and blood.

Domed muscular diaphragm, whose contraction may increase the volume of the thorax.

Emphysema is linked to tobacco smoking, which stimulates the release of proteolytic enzymes from mast cells in the lungs. These enzymes break down alveolar walls producing single large chambers. Thus the effective surface area of the lungs is decreased, causing reduced oxygenation of the blood.

Epiglottis covers trachea when swallowing to prevent entry of food to lungs.

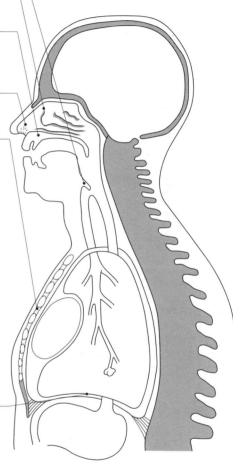

Lung cancer arises from a tumour which develops in the bronchus and then invades adjacent tissues. It causes loss of function and pain, and tumour cells may spread via the bloodstream to other parts of the body.

Pulmonary ventilation is a result of changes in pressure within the thorax.

At rest, when air requirements are low, there is no need for violent breathing movements and the **liver piston** and **diaphragmatic** systems may be sufficient. Heavier demands, for example during exercise, will also require contraction of the external intercostal muscles.

Lungs are passive, elastic structures which are able to 'follow' the movements of the surrounding muscular structures because of the effective negative pressure in the **intrapleural space** between the **pleural membranes.**

Air has a very low density and can therefore respond very quickly to changes in pressure – this permits air-breathing animals to adopt a **tidal system of ventilation.**

The mammalian method involves a **negative pressure system,** i.e. air is **drawn into** the lungs by a fall in pressure within them rather than **forced in** by muscular movements of the mouth. This
a. requires less effort (compare breathing with blowing up a balloon)
b. permits an animal to eat and breathe at the same time (gentle air stream does not pull food downwards).

	INSPIRATION	EXPIRATION
EXTERNAL INTERCOSTAL MUSCLES	Contract – pulling ribs upwards and outwards	Relax – permitting rib cage to move downwards and inwards
DIAPHRAGM	Contracts – moves downwards from domed position	Relaxes – elasticity returns to domed position
LIVER PISTON	Moves downwards	Moves upwards
AIR PRESSURE IN LUNGS	Falls	Increases
AIR MOVEMENT ALONG PRESSURE GRADIENT	Into lungs	Out of lungs
LUNG VOLUME	Increases	Decreases

Rib cage
Diaphragm
Liver piston

An **active** process which decreases lung pressure so that lungs fill by movement of air along a pressure gradient.

A **passive** process which increases lung pressure so that air moves out of lungs along a pressure gradient.

LUNG PRESSURE/kN m⁻²
101

CHANGE IN LUNG VOLUME/cm³
500
0

Fine structure of the lung: exchange of gases in the alveolus requires

1. A tube for the movement of gases to and from the atmosphere (e.g. **bronchiole**).
2. A surface across which gases may be transported between air and blood (i.e. **alveolar membrane**).
3. A vessel which can take away oxygenated blood or deliver carboxylated blood (i.e. a **branch of the pulmonary circulation**).

Inspired air

Expired air

Terminal bronchiole has no rings of cartilage and collapses when external pressure is high – dangerous when diving as trapped air in alveolus may give up nitrogen to blood, where it forms damaging bubbles.

Alveolar duct (atrium)

Alveolus (air sac)

Elastic fibres in alveolus permit optimum extension during inspiration – properties are adversely affected by tobacco smoke → **emphysema.**

Branch of pulmonary artery delivers deoxygenated blood to the alveolar capillaries.

Surfactant is a phospholipid produced by **septal cells** in the alveolar wall. It reduces surface tension of the alveolar walls and prevents them sticking together following expiration - its absence in the newborn may lead to **respiratory distress syndrome,** and even to death, since the effort needed to breathe is 7–10 × normal.

Bronchiole has supporting rings of cartilage to prevent collapse during low pressure phase of breathing cycle.

P₄₅₀ is a cytochrome which speeds oxygen transfer across the alveolar membrane by facilitating diffusion, and is also involved in the detoxification of some harmful compounds by oxidation. Toxins in tobacco smoke may drive P₄₅₀ to **consume** O₂ rather than **transport** it, leading to **anoxia** (oxygen deficiency).

Alveolar-capillary (respiratory) membrane consists of
1. **Alveolar wall:** squamous epithelium and alveolar macrophages.
2. **Epithelial** and **capillary basement membranes.**
3. **Endothelial cells of the capillary wall.**
Despite the number of layers this membrane averages only 0.5 μm in thickness.

Tributary of pulmonary vein returns oxygenated blood to the four pulmonary veins and thence to the left atrium of the heart.

Alveolar capillaries adjacent to the alveolus are the site of oxygen and carbon dioxide transfer between the air in the alveolus (air sac) and the circulating blood.

Stretch receptors provide sensory input, which initiates the **Hering-Breuer reflex** control of the breathing cycle.

Changes in composition of inspired and expired air

	INSPIRED	ALVEOLAR	EXPIRED	
O₂	20.95	13.80	16.40	Oxygen diffuses from alveoli into blood: expired air has an increased proportion of oxygen due to additional oxygen added from the anatomical dead space.
CO₂	0.04	5.50	4.00	Carbon dioxide concentration in alveoli is high because CO₂ diffuses from blood: the apparent fall in CO₂ concentration in expired air is due to dilution in the anatomical dead space.
N₂	79.01	80.70	79.60	The apparent increase in the concentration of nitrogen, a metabolically inert gas, is due to a **relative** decrease in the proportion of oxygen rather than an **absolute** increase in nitrogen.
H₂O(g)	VARI-ABLE		SATURATED	The moisture lining the alveoli evaporates into the alveolar air and is then expired unless the animal has anatomical adaptations to prevent this (e.g. the extensive nasal hairs in desert rats).
Temp.	ATMOS-PHERIC		BODY	Heat lost from the blood in the pulmonary circulation raises the temperature of the alveolar air.

The spirometer and the measurement of respiratory activity

Floating chamber which can be filled with air or oxygen. Chamber moves down when subject inhales and up when subject exhales.

Recording pen which draws a trace on the drum in response to movements of the floating chamber.

Rotating drum which can be set to move at fixed speed.

Scale which can be used to calibrate the apparatus by noting pen movement caused by introduction of known volume of gas.

Mouthpiece with valves ensures one-way flow of air through the apparatus.

INSPIRATION
EXPIRATION

Counterbalance which can be adjusted so that floating chamber moves only in response to breathing movements and is unaffected by gravity.

Lower chamber contains water which seals the system and supports the floating chamber.

Support to hold breathing tubes in fixed position within the floating chamber.

Carbosorb is a solid, non-toxic, carbon dioxide absorbant which can remove CO_2 from exhaled air if the apparatus is used to measure oxygen consumption.

ASTHMA causes **bronchospasm** (narrowing of the bronchioles) and so reduces *peak expiratory flow rate* and *expiratory reserve volume* (↓ by up to 25% in asthma).

PULMONARY VENTILATION (the volume of air available to the lungs per unit of time)

= TIDAL VOLUME
  × BREATHING RATE

P.V. can be increased by increasing either or both tidal volume and breathing rate.

EMPHYSEMA, which destroys the walls of the alveoli, severely reduces the vital capacity (in serious cases the vital capacity may be reduced almost to tidal volume).

Vital capacity = maximum volume of air which can be exchanged from full inspiration to full expiration.

Inspiratory reserve volume = maximum additional volume which can be inhaled after normal tidal inspiration.

Tidal volume = volume of air exchanged during normal quiet breathing.

Expiratory reserve volume = maximum additional volume which can be exhaled following normal tidal expiration.

Functional residual capacity = volume of air which is available for gaseous exchange following tidal expiration.

Residual volume = volume of air which cannot be expelled even during forced expiration: approximately 25 cm³ for each kg body mass.

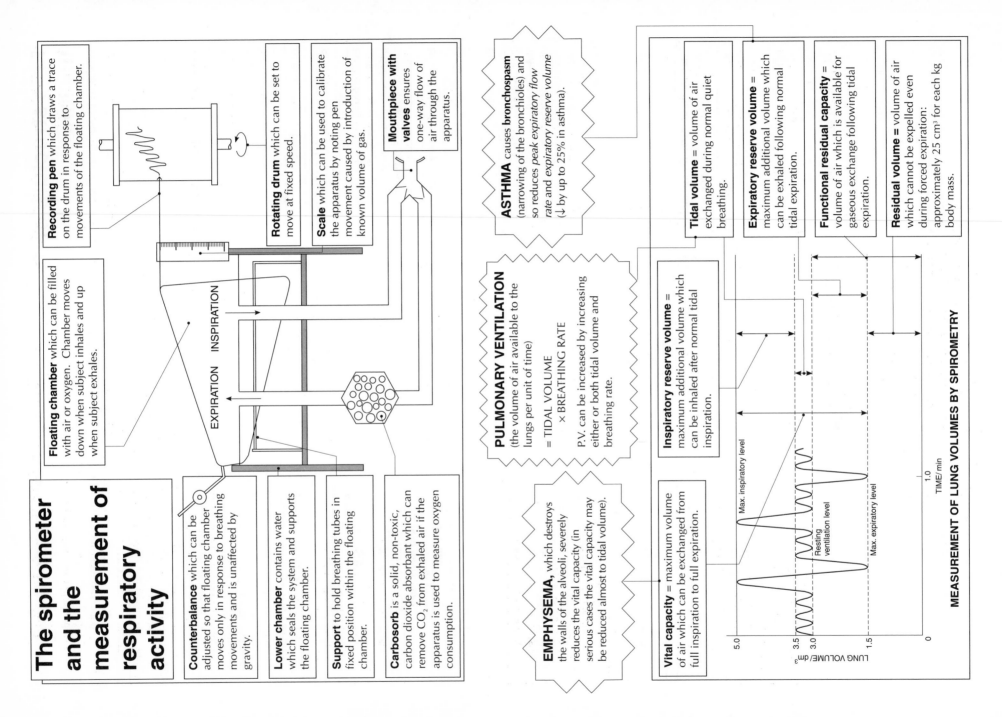

Max. inspiratory level

Resting ventilation level

Max. expiratory level

LUNG VOLUME/dm³

5.0

3.5
3.0

1.5

0

TIME/min

1.0

MEASUREMENT OF LUNG VOLUMES BY SPIROMETRY

Homeostatic control of breathing

involves negative feedback with the medulla acting as an integrator

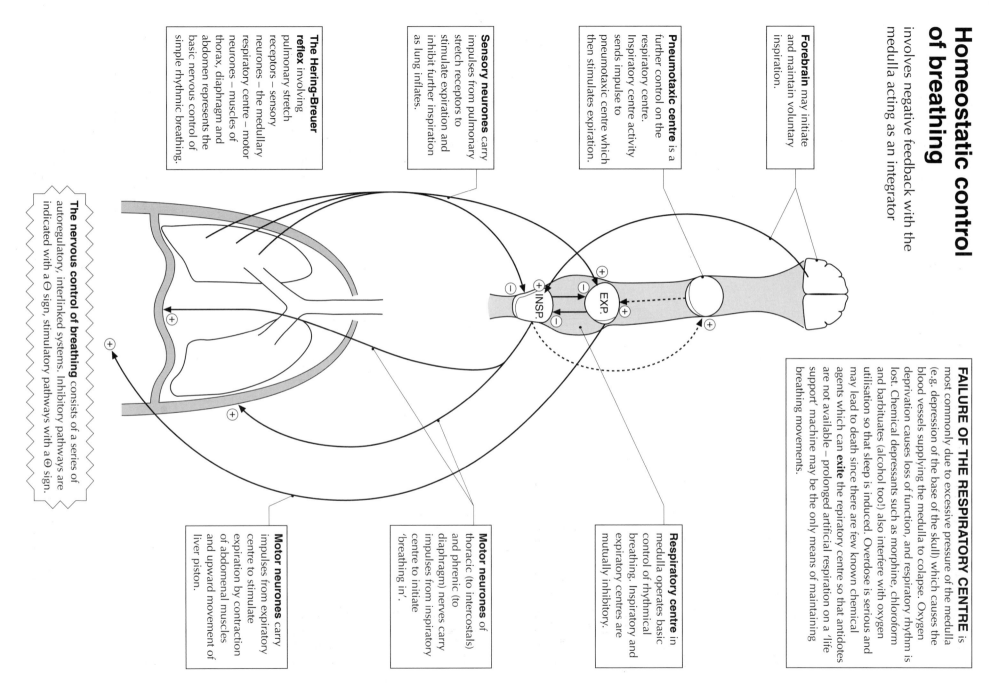

Forebrain may initiate and maintain voluntary inspiration.

Pneumotaxic centre is a further control on the respiratory centre. Inspiratory centre activity sends impulse to pneumotaxic centre which then stimulates expiration.

Sensory neurones carry impulses from pulmonary stretch receptors to stimulate expiration and inhibit further inspiration as lung inflates.

The Hering-Breuer reflex involving pulmonary stretch receptors – sensory neurones – the medullary respiratory centre – motor neurones – muscles of thorax, diaphragm and abdomen represents the basic nervous control of simple rhythmic breathing.

The nervous control of breathing consists of a series of autoregulatory, interlinked systems. Inhibitory pathways are indicated with a ⊖ sign, stimulatory pathways with a ⊕ sign.

Motor neurones carry impulses from expiratory centre to stimulate expiration by contraction of abdomenal muscles and upward movement of liver piston.

Motor neurones of thoracic (to intercostals) and phrenic (to diaphragm) nerves carry impulses from inspiratory centre to initiate 'breathing in'.

Respiratory centre in medulla operates basic control of rhythmical breathing. Inspiratory and expiratory centres are mutually inhibitory.

Mammalian double circulation

comprises *pulmonary* (heart – lung – heart) and *systemic* (heart – rest of body – heart) *circuits*. The complete separation of the two circuits permits rapid, high-pressure distribution of oxygenated blood essential in active, endothermic animals. There are many subdivisions of the systemic circuit, including **coronary, cerebral, hepatic portal** and, during fetal life only, **fetal circuits**. The circuits are typically named for the organ or system which they service – thus each kidney has a **renal** artery and vein. Each organ has an artery bringing oxygenated blood and nutrients, and a vein removing deoxygenated blood and waste.

Pulmonary artery: delivers deoxygenated blood to the lungs for reoxygenation.

Superior vena cava: carries deoxygenated blood back to the right atrium from the head and forelimbs. The venous return to the heart initiates the expansion which triggers the sinu-atrial node to fire the impulse which generates the heart beat.

Inferior vena cava: returns deoxygenated blood at low pressure to the right atrium of the heart.

Hepatic vein: delivers blood with an optimum concentration of solutes (particularly glucose) from the liver to the general circulation.

Hepatic portal vein: transports blood with a very variable solute concentration from the site of solute uptake (the gut) to the site of storage or regulation (the liver).

Renal veins: return deoxygenated blood with a reduced concentration of urea and creatinine, and a regulated pH and Na$^+$/K$^+$ ratio, from the kidneys to the general circulation.

Pulmonary veins: return oxygenated blood at low pressure to the left atrium of the heart.

Carotid arteries (to head) give pulse over temporal bone of skull. Walls contain carotid chemoreceptors (CO_2 and O_2) and baroreceptors (pressure) which transmit information to integration centres in brain.

Aorta: the principal vessel which distributes oxygenated blood at high pressure to the systemic circulation.

Brachial arteries (to arm) are used by physicians to monitor blood pressure.

Hepatic artery: delivers oxygenated blood to the liver - this organ is so active metabolically that oxygen demands are very high.

Mesenteric artery: delivers oxygenated blood to the gut – demand for muscle contraction is low, but active transport mechanisms require ATP generated by (preferably aerobic) respiration.

Renal artery: carries blood with high O_2 concentration and high concentrations of solutes such as urea, creatinine and Na$^+$.

All arteries have a thick, muscular elastic wall but contain no valves. The **pulse** can be felt where an artery crosses a bone.

PULMONARY CIRCULATION

SYSTEMIC CIRCULATION

LUNGS

RIGHT ATRIUM

RIGHT VENTRICLE

LEFT ATRIUM

LEFT VENTRICLE

LIVER

INTESTINES

KIDNEYS

OTHER ORGANS

HEAD AND ARMS

Veins have thin, non-muscular walls, but contain **semi-lunar valves** to prevent backflow of low pressure blood.

Capillaries are present in all organs/tissues and are the sites of exchanges between blood and tissue fluid.

The **flow of blood** is maintained in three ways.

1. **The pumping action of the heart:** the ventricles generate pressures great enough to drive blood through the arteries into the capillaries.

2. **Contraction of skeletal muscle:** the contraction of muscles during normal movements compress and relax the thin-walled veins causing pressure changes within them. Pocket valves in the veins ensure that this pressure directs the blood to the heart, without backflow.

3. **Inspiratory movements:** reducing thoracic pressure caused by chest and diaphragm movements during inspiration helps to draw blood back towards the heart.

Blood cells differ in structure and function.

If blood is spun for a few minutes in a high speed centrifuge it separates into two layers.

Serum is the name given to plasma from which the soluble protein fibrinogen (a protein involved in blood clotting) has been removed (as fibrin).

PLASMA (55%)

CELLS (45%)

Blood cells originate from **stem cells** in the bone marrow by the process of **haemopoiesis.**

UNCOMMITTED STEM CELL

MONOBLAST HAEMOCYTOBLAST LYMPHOBLAST

MEGA KARYOCYTE

PROERYTHROBLAST

MYELOBLAST

ERYTHROCYTE LEUCOCYTE PLATELETS
(e.g. neutrophil)

LYMPHOCYTE

MACROPHAGE

SCALE
5 µm

Eosinophils have a double-lobed nucleus and granules which stain red with the acid dye eosin. They help control the allergic response – for example, they secrete enzymes which inactivate histamine. Their numbers increase during allergic reactions and in response to some parasitic infections (e.g. tapeworm and hookworm).

Erythrocytes (red blood cells) are the most numerous of blood cells – about 5 000 000 per mm³ of adult blood. They function in the transport of O_2 and CO_2, and they contribute to the buffering capacity of the blood. The red colour is due to the presence of the pigment haemoglobin. There are several advantages in packing the haemoglobin into cells rather than leaving it free in the cytoplasm – it keeps the blood viscosity low, it allows the best arrangement of enzymes and solutes for functioning of haemoglobin and it prevents a dramatic reduction in blood water potential. The typical lifespan of a red cell is 90–120 days, before they are destroyed in the spleen.

Platelets (thrombocytes) are fragments of cells which are involved in blood clotting (they disintegrate to release thromboplasts).

Neutrophils are the most abundant of the leucocytes (white blood cells). They are very short-lived (12–72 h), contain non-staining granules, and are responsible for the phagocytosis of micro-organisms. They migrate from the blood to the tissues, and are so active in phagocytosis that they are replaced at the rate of about 100 000 000 000 per day.

Monocytes are the largest of the leucocytes. They are agranulocytes (have non-granular cytoplasm) and have a large, bean-shaped nucleus. They spend a short time (2–3 days) in the circulatory system before moving into the tissues where they mature into phagocytic macrophages.

Lymphocytes make up about 30% of the circulating leucocytes. Although they are produced in the bone marrow they continue to develop and mature in the lymph nodes, the thymus gland and the spleen. They are responsible for the specific immune response – the B-lymphocytes produce antibodies and the T-lymphocytes have a number of roles, including co-ordination of the immune response and direct cell destruction. They are best identified by the prominent, deeply staining nucleus and the thin 'halo' of clear cytoplasm.

Basophils have an S-shaped nucleus and granules which stain blue. They secrete large amounts of histamine (which increases inflammation) and heparin (which helps to keep a balance between blood clotting and not clotting).

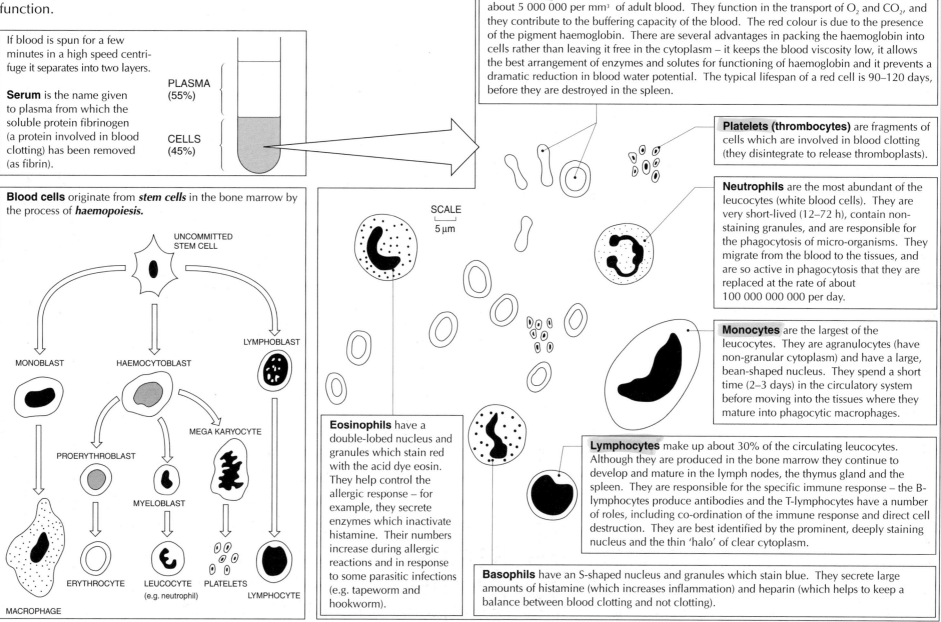

Haemoglobin – oxygen association curve

shows the relationship between haemoglobin saturation with oxygen (i.e. the % of haemoglobin in the form of oxyhaemoglobin) and the partial pressure of oxygen in the environment (the pO_2 or oxygen tension). The curve may be defined by the LOADING and UNLOADING TENSION.

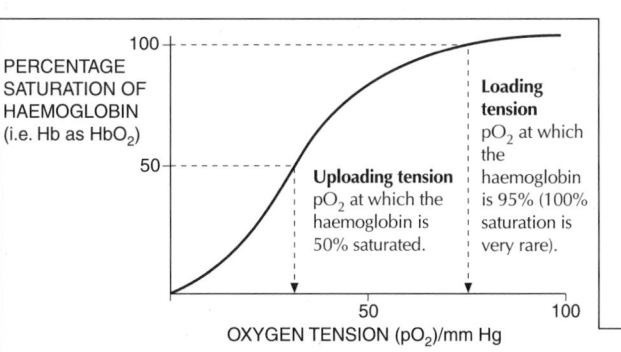

PERCENTAGE SATURATION OF HAEMOGLOBIN (i.e. Hb as HbO_2)

Uploading tension pO_2 at which the haemoglobin is 50% saturated.

Loading tension pO_2 at which the haemoglobin is 95% (100% saturation is very rare).

OXYGEN TENSION (pO_2)/mm Hg

The S-shape of the curve is most significant. Simply put, it means that oxygen associates with haemoglobin, and remains associated with it, at oxygen tensions typical of the alveolar capillaries, the pulmonary vein, the aorta and the arteries, but that it very rapidly dissociates from haemoglobin at oxygen tensions typical of those found in respiring tissues. This dissociation is almost complete at the low oxygen tensions found in the most active tissues. In other words, oxygen release from oxyhaemoglobin is tailored to the tissues' demands for this gas.

The reasons for this S-shapedness is that haemoglobin and oxygen illustrate co-operative binding, that is, the binding of the first oxygen molecule to haemoglobin alters the shape of the haemoglobin molecule slightly so that the binding of a second molecule of oxygen is made easier, and so on until haemoglobin has its full complement of four molecules of oxygen. Conversely, when one molecule of oxygen dissociates from the oxyhaemoglobin, the haemoglobin, shape is adjusted to make release of successive molecules of oxygen increasingly easy.

BOHR SHIFT

We have noted that the release of oxygen from oxyhaemoglobin is well suited to the oxygen demands of the tissues. The balance between supply and demand is even further adjusted by special local conditions in the tissues. For example, the curve is shifted to the right and made steeper by higher tensions of carbon dioxide, by falling pH, by increasing lactate concentration and by rising temperatures. All of these changes correspond to increased respiratory activity in a tissue, and the alteration in the oxygen association curve ensures that oxygen unloads from oxyhaemoglobin even more rapidly. This phenomenon is called the Bohr shift.

A similar Bohr shift of the Hb – O_2 association curve **to the right** (i.e. towards more O_2 released) results from **falling pH, increased [lactate], increased temperature.**

Haemoglobin and DPG: **Diphosphoglycerate** is found in erythrocytes during glycolysis, and is thus a good indicator of metabolic rate in cells. It binds reversibly to haemoglobin and promotes the release of oxygen from oxyhaemoglobin

The **'difference'** between the two curves represents the additional oxygen released from oxyhaemoglobin at any particular pO_2 (here shown as T mm Hg)

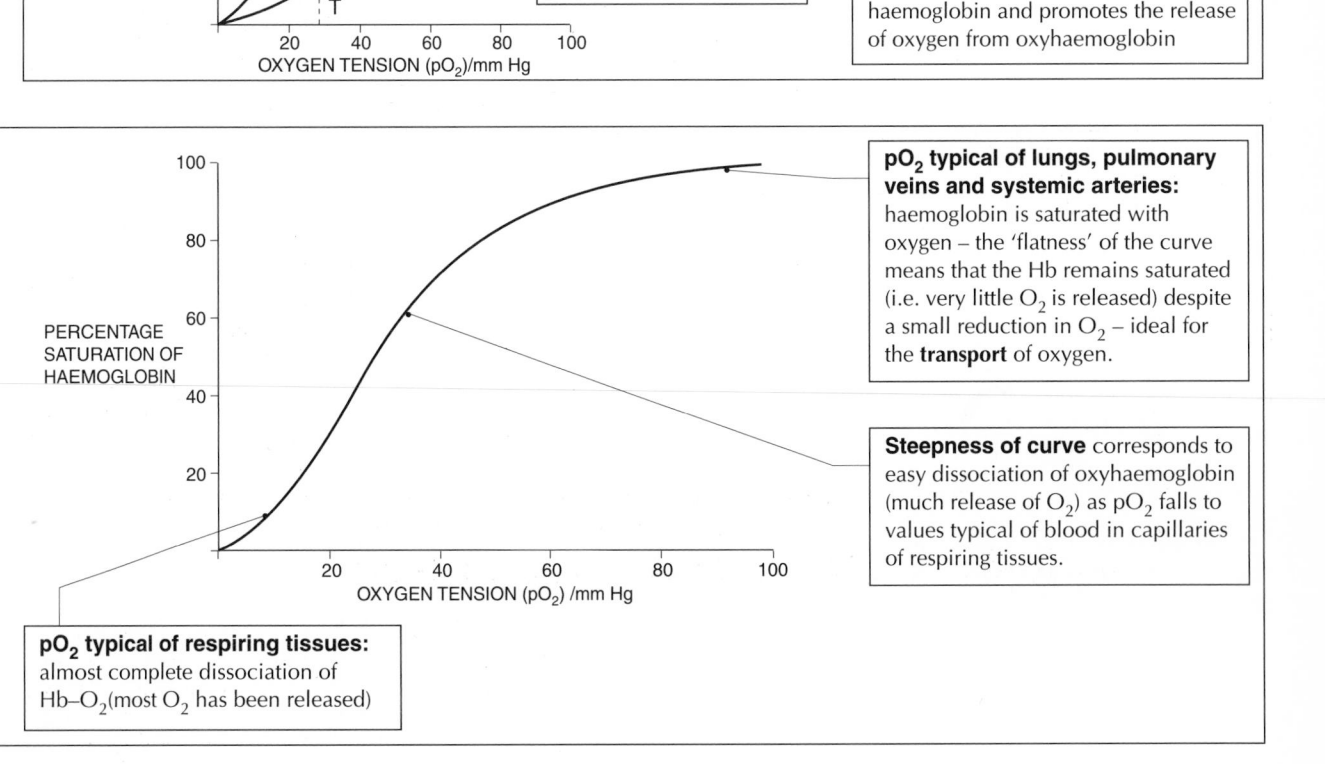

pO_2 typical of lungs, pulmonary veins and systemic arteries: haemoglobin is saturated with oxygen – the 'flatness' of the curve means that the Hb remains saturated (i.e. very little O_2 is released) despite a small reduction in O_2 – ideal for the **transport** of oxygen.

Steepness of curve corresponds to easy dissociation of oxyhaemoglobin (much release of O_2) as pO_2 falls to values typical of blood in capillaries of respiring tissues.

pO_2 typical of respiring tissues: almost complete dissociation of Hb–O_2 (most O_2 has been released)

MYOGLOBIN, MUSCLE AND MARATHONS

The muscles of mammals contains a red pigment called myoglobin which is structurally similar to one of the four sub-units of haemoglobin. This pigment may also bind to oxygen, but since there is only one haem group there can be no co-operative binding and the myoglobin-oxygen association curve is hyperbolic rather than sigmoidal.

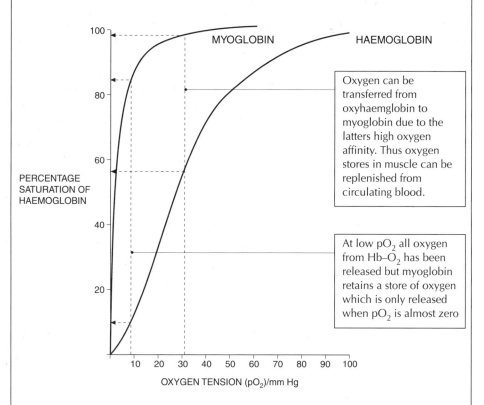

Oxygen can be transferred from oxyhaemglobin to myoglobin due to the latters high oxygen affinity. Thus oxygen stores in muscle can be replenished from circulating blood.

At low pO_2 all oxygen from $Hb–O_2$ has been released but myoglobin retains a store of oxygen which is only released when pO_2 is almost zero

The significance of this is that at any particular oxygen tension myoglobin has a higher affinity for oxygen than does haemoglobin. Thus when oxyhaemoglobin in blood passes through tissues such as muscle, oxygen is transferred to the myoglobin. Further analysis of the myoglobin-oxygen association curve will show that myoglobin does not release oxygen until oxygen tension is very low indeed, and myoglobin therefore represents an excellent store for oxygen.

Muscles which have a high oxygen demand during exercise, or which may be exposed to low oxygen tension in the circulating blood, commonly have particularly large myoglobin stores and are called 'red' muscles. A high proportion of red muscle is of considerable advantage to marathon runners who must continue to respire efficiently even when their blood is severely oxygen-depleted.

Myoglobin and foetal haemoglobin

have molecular properties which promote oxygen transfer from circulating blood to skeletal muscle and to the foetus

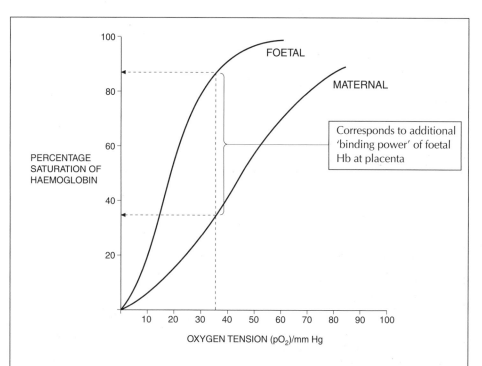

Corresponds to additional 'binding power' of foetal Hb at placenta

FOETAL HAEMOGLOBIN IN MAMMALS has a structure which differs slightly from maternal haemoglobin. As a result the foetal form has a higher affinity for oxygen and the gas is readily transferred across the placenta from mother to foetus. As birth approaches foetal haemoglobin is gradually replaced by the adult form.

At the respiring tissue: carbon dioxide produced in the mitochondria diffuses out of the cells, through the plasma, and into the erythrocytes, where it combines with water to produce carbonic acid, H_2CO_3, under the influence of the enzyme carbonic anhydrase.
①

The hydrogencarbonate concentration in the red cells falls, more diffuses in from the plasma, and the process continues so that more carbon dioxide is released. Once more electrical neutrality of the erythrocytes is maintained by the chloride shift, but this time the chloride ions are moving in the opposite direction, from the cell to the plasma.
⑥

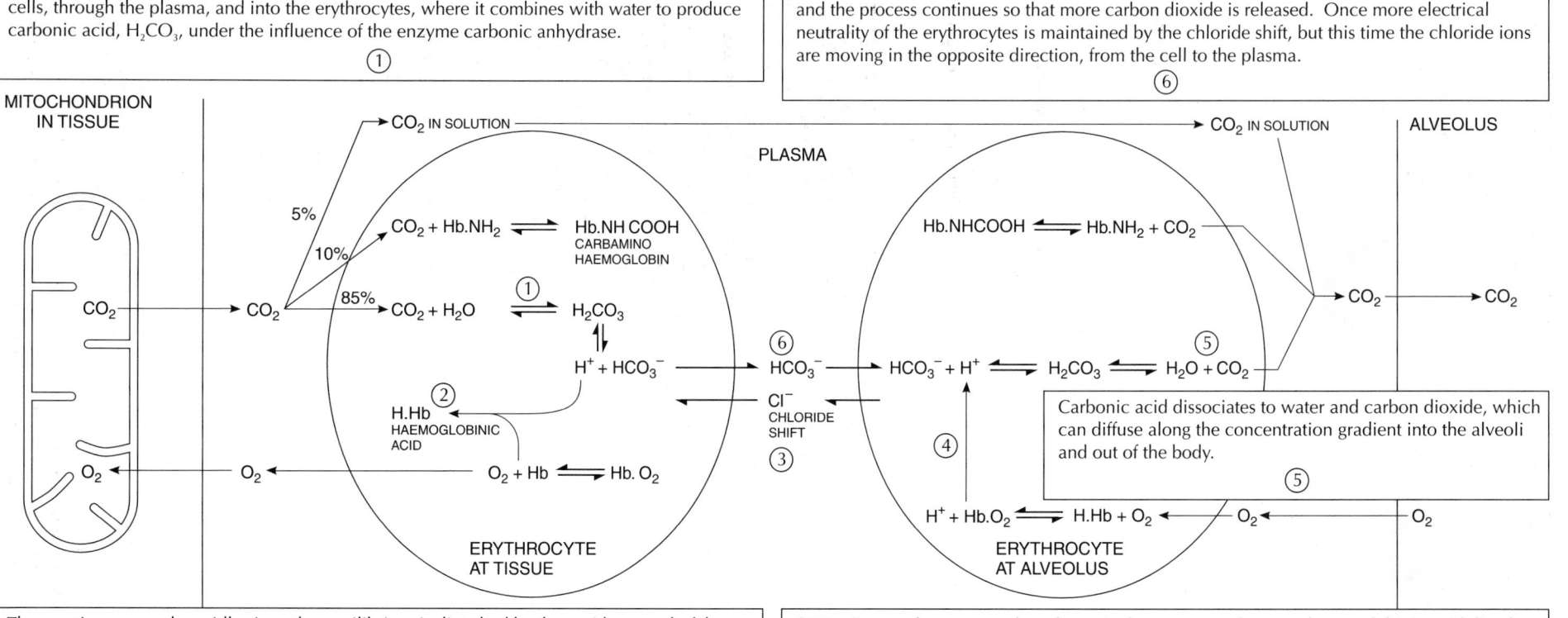

MITOCHONDRION IN TISSUE

PLASMA

ALVEOLUS

CO_2 IN SOLUTION

CO_2 IN SOLUTION

CO_2

CO_2

5%

10%

85%

$CO_2 + Hb.NH_2 \rightleftharpoons Hb.NH\,COOH$
CARBAMINO HAEMOGLOBIN

Hb.NHCOOH $\rightleftharpoons$ Hb.NH$_2$ + CO$_2$

$CO_2 + H_2O$ ① H_2CO_3

$H^+ + HCO_3^-$ ⑥ HCO_3^-

$HCO_3^- + H^+ \rightleftharpoons H_2CO_3 \rightleftharpoons H_2O + CO_2$ ⑤

CO_2

CO_2

② H.Hb
HAEMOGLOBINIC ACID

Cl$^-$ CHLORIDE SHIFT ③

O_2

O_2

$O_2 + Hb \rightleftharpoons Hb.O_2$

④

$H^+ + Hb.O_2 \rightleftharpoons H.Hb + O_2$

O_2

O_2

ERYTHROCYTE AT TISSUE

ERYTHROCYTE AT ALVEOLUS

Carbonic acid dissociates to water and carbon dioxide, which can diffuse along the concentration gradient into the alveoli and out of the body.
⑤

The reaction proceeds rapidly since the equilibrium is disturbed by the rapid removal of the hydrogen ions (H^+) by association with haemoglobin to form haemoglobinic acid (H.Hb). By accepting hydrogen ions in this way haemoglobin is acting as a buffer, permitting the transport of large quantities of carbon dioxide without any significant change in blood pH.
②

At the lungs: the reverse takes place. In the presence of oxygen haemoglobinic acid dissolves so that oxyhaemoglobin (Hb.O$_2$) may be formed. This releases hydrogen ions which combine with hydrogencarbonate in the plasma to produce carbonic acid.
④

The transport of carbon dioxide from tissue to lung

The red cell and haemoglobin both play a significant part in this process as well as in the transport of oxygen.

As a result of these changes the hydrogencarbonate concentration in the erythrocyte rises and these ions begin to diffuse along a concentration gradient into the plasma. However, this movement of negative ions is not balanced by an equivalent outward flow of positive ions since the membrane of the erythrocyte is relatively impermeable to sodium and potassium ions, which are therefore retained within the cell. This could potentially be disastrous, since positively charged erythrocytes would repel one another, a situation which would not enhance their function as oxygen carriers in the confines of a closed circulatory system! The situation is avoided, and electrical neutrality maintained, by an inward diffusion of chloride ions from the plasma sufficient to balance the HCO$_3^-$ moving out. This movement of chloride ions to maintain erythrocyte neutrality is called the **chloride shift.**
③

Tissue fluid (interstitial or intercellular fluid)

fluid) is the immediate environment of the cells, and represents the 'internal environment' described by Claude Bernard in his definition of homeostasis.

Capillary wall is one cell thick

ARTERIAL END OF CAPILLARY

Plasma proteins do not move from plasma to tissue fluid (cannot cross capillary endothelium) - largely responsible for solute potential of plasma.

Living cells place demands on the tissue fluid.

Movement from plasma to tissue fluid
- Water
- Oxygen
- Soluble products of digestion
- Hormones

Movement from tissue fluid to plasma
- Water
- Carbon dioxide
- Nitrogenous waste
- Hormones and other secretions

VENOUS END OF CAPILLARY

FORCES WHICH REGULATE THE FORMATION AND RECLAMATION OF TISSUE FLUID

Net force driving fluid movement at any point = pressure potential gradient – solute potential gradient

Pressure potential (hydrostatic potential) is the pressure exerted on a fluid by its surroundings, e.g. by **pumping action of heart** and **elastic recoil of arteries.**

Solute potential (osmotic potential) is the force of attraction towards water molecules caused by dissolved solutes, particularly **ions** and **plasma proteins.**

Venous end of capillaries: the PP gradient has fallen (1) as distance from pumping heart increases and (2) as volume of fluid in vessels falls. High concentration of plasma proteins means that blood solute potential is high.

SP GRADIENT > PP GRADIENT

Net movement of water from tissues to plasma.

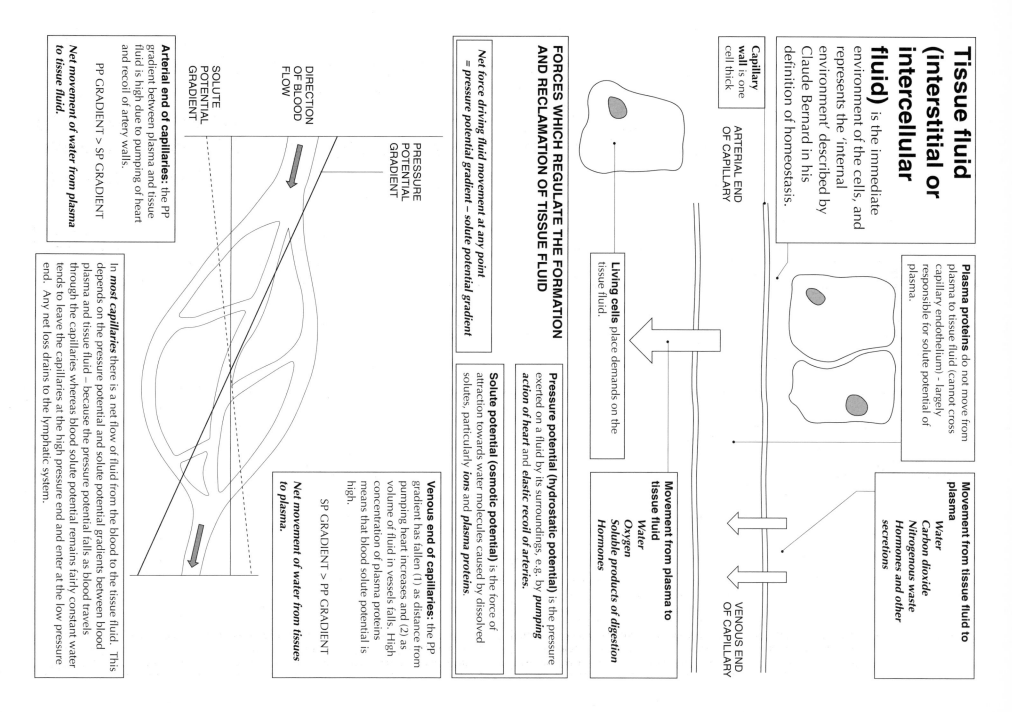

DIRECTION OF BLOOD FLOW

PRESSURE POTENTIAL GRADIENT

SOLUTE POTENTIAL GRADIENT

Arterial end of capillaries: the PP gradient between plasma and tissue fluid is high due to pumping of heart and recoil of artery walls.

PP GRADIENT > SP GRADIENT

Net movement of water from plasma to tissue fluid.

In **most capillaries** there is a net flow of fluid from the blood to the tissue fluid. This depends on the pressure potential and solute potential gradients between blood plasma and tissue fluid – because the pressure potential falls as blood travels through the capillaries whereas blood solute potential remains fairly constant water tends to leave the capillaries at the high pressure end and enter at the low pressure end. Any net loss drains to the lymphatic system.

Mammalian heart: structure and function

Pulmonary (semilunar) valve: is composed of three cusps or watchpocket flaps which are forced together then the pressure in the pulmonary artery exceeds that in the right ventricle, thus preventing backflow of blood into the relaxing chambers of the heart.

P ventricle > *P* artery

P ventricle < *P* artery

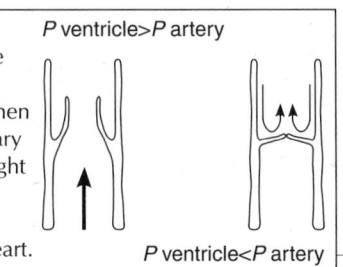

Tricuspid (right atrioventricular) valve: has three fibrous flaps with pointed ends which point into the ventricle. The flaps are pushed together when the ventricular pressure exceeds the atrial pressure so that blood is propelled past the inner edge of the valve through the pulmonary artery instead of through the valve and back into the atrium.

Superior (anterior) vena cava: carries deoxygenated blood back to the right atrium of the heart. As with other veins the wall is thin, with little elastic tissue or smooth muscle. In contrast to veins returning blood from below the heart there are no venous valves, since blood may return under the influence of gravity.

Control of heartbeat

1. The heartbeat is initiated in the **sinu-atrial node** (particularly excitable myogenic tissue in the wall of the right atrium).
2. **Intrinsic** heart rate is about 78 beats per minute.
3. **External (extrinsic)** factors may modify basic heart rate:
   a. **vagus nerve** decreases heart rate;
   b. **accelerator (sympathetic) nerve** increases heart rate;
   c. **adrenaline** and **thyroxine** increase heart rate.

Resting heart rate of 70 beats per minute indicates that heart has **vagal tone.**

Drugs such as **beta-blockers** prevent the binding of adrenaline to β-**receptors** and so may be prescribed to control heartbeat during times of stress, e.g. examinations.

Aorta: carries oxygenated blood from the left ventricle to the systemic circulation. It is a typical elastic (conducting) artery with a wall that is relatively thick in comparison to the lumen, and with more elastic fibres than smooth muscle. This allows the wall of the aorta to accommodate the surges of blood associated with the alternative contraction and relaxation of the heart - as the ventricles contract the artery expands and as the ventricle relaxes the elastic recoil of the artery forces the blood onwards.

Pulmonary arteries

Left atrium

Right atrium

Right ventricle: generates pressure to pump deoxygenated blood to pulmonary circulation.

Myocardium is composed of cardiac muscle: intercalated discs separate muscle fibres, strengthen the muscle tissue and aid impulse conduction; cross-bridges promote rapid conduction throughout entire myocardium; numerous mitochondria permit rapid aerobic respiration. Cardiac muscle is myogenic (can generate its own excitatory impulse) and has a long refractory period (interval between two consecutive effective excitatory impulses), which eliminates danger of cardiac fatigue.

The pressure generated by the left ventricle is greater than that generated by the right ventricle as the systemic circuit is more extensive than the pulmonary circuit.

The pressure generated by the atria is less than that generated by the ventricles since the distance from atria to ventricles is less than that from ventricles to circulatory system.

Volume: the same volume of blood passes through each side of the heart, so circulating volumes are also equal in pulmonary and systemic circuits.

Aortic (semilunar) valve: prevents backflow from aorta to left ventricle.

Bicuspid (mitral, left atrioventricular) valve: ensures blood flow from left ventricle into aortic arch.

Left ventricle: generates pressure to force blood into the systemic circulation.

Chordae tendineae: short, inextensible fibres – mainly composed of collagen – which connect to free edges of atrioventricular valves to prevent 'blow-back' of valves when ventricular pressure rises during contraction of myocardium.

Papillary muscles: contract as wave of excitation spreads through ventricular myocardium and tighten the chordae tendineae just before the ventricles contract.

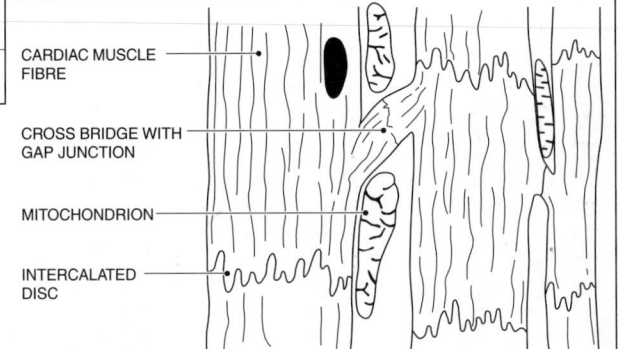

CARDIAC MUSCLE FIBRE

CROSS BRIDGE WITH GAP JUNCTION

MITOCHONDRION

INTERCALATED DISC

Blood flow through the heart

is controlled by two phenomena: the opening and closing of the valves and the contraction and relaxation of the myocardium. Both activities occur without direct stimulation from the nervous system; the valves are controlled by the pressure changes in each heart chamber, and the contraction of the cardiac muscle is stimulated by its conduction system.

Impulse transmission through the heart's conduction system generates electrical currents which can be recorded using an electrocardiograph and presented as an **electrocardiogram** (ECG).

The P wave indicates **atrial depolarisation** – the spread of an impulse from the SA node through the two atria. It is less powerful than the QRS complex since the atria are less massive than the ventricles.

As ventricular pressure exceeds atrial pressure the **semilunar values** open to allow blood from ventricles to arteries

Contraction of the ventricles without any emptying (note no reduction in volume) causes a sharp rise in ventricular pressure

Sharp rise in ventricular pressure above atrial pressure closes the **cuspid valves** to prevent backflow of blood from ventricles to atria

(c) VOLUME OF LEFT VENTRICLE (cm³)

Ventricular volume rises as blood moves (passively and by atrial contraction) from atria

Ventricular volume falls as ventricles contract and semilunar valves open allowing blood flow through to arteries

HEART SOUNDS are caused by turbulence in blood flow created by closure of the valves. The first sound ("lubb") is caused by the closure of the cuspid (atrioventricular) valves soon after ventricular systole begins, the second sound ("dupp") is created by the closure of the semilunar valves towards the end of ventricular systole. The pause between the second sound and the first sound of the next cycle is about twice as long as the pause between the first and second sound of each cycle. Thus the cardiac cycle can be heard as "lubb, dupp, pause; lubb, dupp, pause; lubb, dupp, pause".

Blood flows from an area of higher pressure to one of lower pressure. The diagram relates pressure changes to heartbeat in the left side of the heart – the same pattern exists in the right side although pressures are somewhat lower there.

The QRS wave (complex) represents **ventricular depolarisation** and is strong enough to mask atrial repolarisation.

The T wave represents **ventricular repolarisation.**

The dicrotic notch is the slight increase in pressure due to recoil of the arteries as the semilunar valve closes

As the ventricular pressure falls below the arterial pressure the **semilunar valves** snap shut to prevent the backflow of blood from arteries to ventricle.

The arterial blood pressure falls as the ventricles relax (no blood entering arteries) and blood moves along systemic and pulmonary circuits.

About 70% of blood flows passively from atria to ventricles but this increase in pressure during atrial systole is necessary to force the remaining 30% through to the ventricles.

As the atrial pressure rises due to atrial filling the ventricular pressure falls as the ventricular muscle relaxes. As the ventricular pressure falls below the atrial pressure the **cuspid valves** open allowing blood to flow (initially passively) from atria to ventricles.

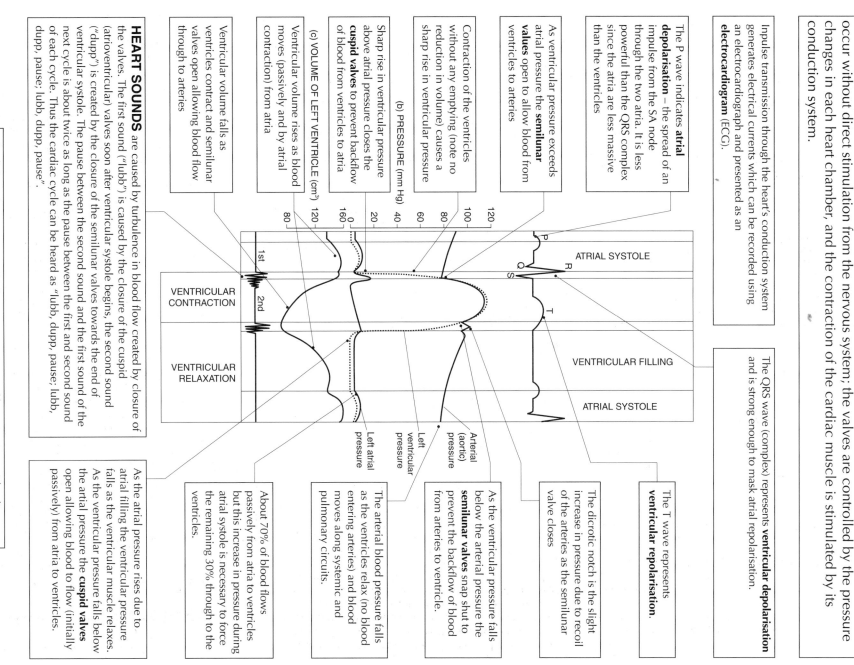

(b) PRESSURE (mm Hg)

120
100
80
60
40
20
0
160
120
80

1st
2nd

P
Q
R
S
T

ATRIAL SYSTOLE

VENTRICULAR CONTRACTION

VENTRICULAR RELAXATION

VENTRICULAR FILLING

ATRIAL SYSTOLE

Arterial (aortic) pressure

Left ventricular pressure

Left atrial pressure

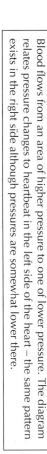

Heart disease

affects 1 in 4 persons between the ages of 30 and 60.

High risk factors include
1 High blood cholesterol level
2 Cigarette smoking
3 Obesity
4 Stress
5 Lack of regular exercise

Pericarditis and myocarditis are inflammation of the pericardium and myocardium respectively. Such inflammations may result from bacterial infection (e.g. following a tooth extraction or respiratory complaint) or a depressed immune response following chemotherapy or radiotherapy.

Valvular disease: Stenosis is caused by fibrosis of the valve following valve inflammation, **valvular incompetence** is a functional defect caused by failure of a valve to close completely, allowing blood to flow back into a ventricle as it relaxes.

Ishaemic heart disease is due to narrowing or closure of one or more branches of the coronary arteries. The narrowing is caused by atheromatous plaques, closure either by plaques alone or complicated by thrombosis.

Angina pectoris is severe ischaemic pain following physical effort. The narrowing vessel may supply sufficient oxygen and nutrient laden blood to meet the demands of the cardiac muscle during rest or moderate exercise but is unable to dilate enough to allow for the increased blood flow needed by an active myocardium during severe exercise. Angina is treated by a combination of analgesics (painkillers) and vasodilators.

Congenital defects:

Coarction of the aorta – one segment of the aorta is too narrow so that the flow of oxygenated blood to the body is reduced.

Persistent ductus arteriosus – the ductus arteriosus should close at birth – if it does not blood flows between the aorta and the pulmonary artery, reducing the volume entering the **systemic** circulation but causing congestion in the **pulmonary** circulation.

Septal defect ("Hole in the heart") is most commonly due to incomplete closure of the foramen ovale at birth, allowing blood to flow between the two atria. This may reduce the flow of oxygenated blood or more severely may raise right sided blood pressure to dangerous levels.

Rheumatic fever (rarely seen today) is an autoimmune disease which occurs 2–4 weeks after a throat infection, caused by *Streptococcus pyogenes*. The antibodies developed to combat the infection cause damage to the heart, in a way that is not yet understood.

Effects include
1 development of fibrous nodules on the mitral and aortic valves (rarely on the tricuspid and pulmonary valves)
2 fibrosis of the connective tissue of the mycardium
3 accumulation of fluid in the pericardial cavity – in severe cases the cavity is obliterated and heart expansion during diastole is severely affected

---

Myocardial infarction (commonly called "heart attack"): Infarction means the death of an area of tissue following an interrupted blood supply. The tissue beyond the obstruction dies and is replaced by non-contractile tissue – the heart loses some of its strength. In CHRONIC ISCHAEMIC HEART DISEASE many small infarcts form and collectively and gradually lead to myocardial weakness, but in ACUTE ISCHAEMIC HEART DISEASE the infarct is large since a large artery is closed – death may occur suddenly for a number of reasons:

1 Acute cardiac failure due to shock
2 Severe arrhythmia due to disruption of the conducting system
3 Rupture of the ventricle wall, usually within two weeks of the infarction

Recent studies suggest that much tissue damage occurs when blood flow is **restored** not when it is **interrupted.** When blood supply is interrupted the affected tissue releases an enzyme called XANTHOXIDASE – when blood supply is restored the enzyme reacts with oxygen in the blood to produce SUPEROXIDE FREE RADICALS which cause enormous tissue damage by protein oxidation. Drugs such as SUPEROXIDE DISMUTASE and ALLOPURINOL, are used to prevent the formation of the free radicals and limit tissue damage.

Disorders of the circulation

Aneurysms are thin, weakened sections of the wall of an artery or a vein that bulges outward forming a balloon-like sac. Aneurysms may be congenital or may be caused by atherosclerosis, syphilis or trauma. The vessel may become so thin that it bursts causing massive haemorrhage, pain and severe tissue damage. Unruptured aneurysms may exert pressure on adjacent tissues e.g. aortic aneurysm may compress the oesophagus, causing difficulty in swallowing.

Hypertension (high blood pressure) is the most common circulatory disease. Normal blood pressure is $^{120}/_{80}$ – any value above $^{140}/_{90}$ indicates hypertension and $^{160}/_{95}$ is considered dangerous. Uncontrolled hypertension may damage heart (heart enlarges and demands more oxygen), brain (CVA, or stroke, which ruptures cerebral arteries supplying brain) and kidneys (glomoular arterioles are narrowed and deliver less blood). Treatment includes weight reduction in obese patients, restriction of Na^{2+} intake, increased Ca^{2+}/K^+ intake and stopping smoking. Vasolilators, beta blockers and diuretics are widely prescribed.

Varicose veins are so dilated that the valves do not close to prevent backflow of blood. The veins lose their elasticity and become congested. Predisposing factors include age, heredity, obesity, gravity and compression by adjacent structures. Common sites are the legs (where poor circulation may lead to haemorrhage or ulceration of poorly-nourished skin), the rectal-anal junction (haemorrhoids may bleed and cause mild anaemia), the oesophagus (caused by liver cirrhosis or right-sided heart failure – may lead to haemorrhage and death) and the spermatic cord (may reduce spermatogenesis).

Embolism is obstruction of a blood vessel by an abnormal mass of material (an embolus) such as a fragment of blood clot, atheromatous plaque, bone fragment or air bubble. Emboli may lodge in the coronary vessels, the lungs or the liver causing restricted circulation and, perhaps, death.

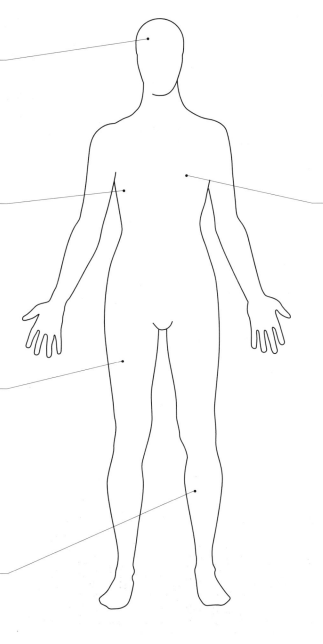

Atherosclerosis is a process in which fatty substances are deposited in the walls of arteries in response to certain stimuli. One possible sequence is

CO from smoking and/or hypertension and/or high available cholesterol ⇒ damage to arterial endothelium ⇒ invasion of arterial wall by monocytes ⇒ monocytes develop into macrophages ⇒ macrophages take up cholesterol from low density lipoproteins **+** smooth muscle cells ingest cholesterol ⇒ ATHEROMA OR ATHEROSCLEROTIC PLAQUE

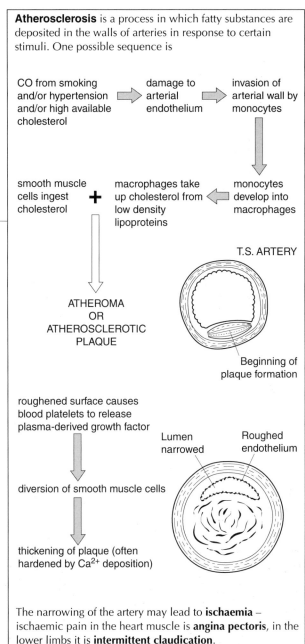

T.S. ARTERY

Beginning of plaque formation

roughened surface causes blood platelets to release plasma-derived growth factor ⇒ diversion of smooth muscle cells ⇒ thickening of plaque (often hardened by Ca^{2+} deposition)

Lumen narrowed

Roughed endothelium

The narrowing of the artery may lead to **ischaemia** – ischaemic pain in the heart muscle is **angina pectoris**, in the lower limbs it is **intermittent claudication**.

Vasomotor centre is the **integrator** in the control of circulatory changes. It is located in the **medulla** and has **sensory input** from cerebral cortex and baroreceptors and **motor output** to heart, precapillary sphincters and adrenal glands.

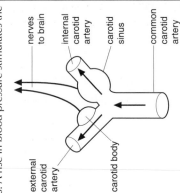

nerves to brain

internal carotid artery

carotid sinus

common carotid artery

external carotid artery

carotid body

Changes in blood pressure: at the base of the *internal* carotid artery there is a swelling, the carotid sinus, that contains stretch receptors. A rise in blood pressure stimulates the stretch receptors and sensory impulses pass to the cardiovascular centre in the brain. Motor impulses pass from the brain to the heart and peripheral arterioles which cause a decrease in heart rate and vasoconstriction. Thus the blood pressure falls again.

Higher centres in the brain can also have an effect on the circulatory system. During periods of excitement (waiting for an examination) or shock (getting the exam results), the cerebral cortex passes nerve impulses to the adrenal glands which respond by secreting the hormone adrenaline into the bloodstream. Adrenaline increases cardiac frequency and constriction of the peripheral arterioles, an effect similar to increasing blood pressure by stimulation from the cardiovascular centre. Although the nervous and hormonal systems produce identical effects, there is an important difference. The nervous system (cardiovascular centre) adjusts the body to changes **that are taking place**. The hormonal system (cerebral cortex/adrenaline) adjusts the body to changes **that have yet to occur**. The hormonal system of control prepares the body for action in emergencies.

Changes in cardiac output: the secretion of adrenaline from the adrenal medulla and the activity of the sympathetic nervous system both increase cardiac output. This is a result of both increased heart rate and increased stroke volume.

The circulatory system in action:
adjustment to exercise

The circulatory system must be able to respond to the changing requirements of the tissues which it supplies. The circulatory system does not work in isolation, and changes in circulation are often accompanied by adjustments to breathing rate and skeletal muscle tone.

The response to exercise involves
1. A change in cardiac output
2. A change in blood distribution
3. A change in blood pressure
4. A change in gas exchange

Note the typical homeostatic sequence

STIMULUS

↓

RECEPTOR

↓

INTEGRATOR

↓

EFFECTOR

↓

RESPONSE

↓

RETURN TO NORM

Local control of blood flow

There is also a local control of blood flow and distribution since the sphincters present at the entrances to capillary beds and arterio-venous shunts are sensitive to local changes in the concentration of CO_2 and lactate. Thus during exercise an increase in these waste products causes relaxation of the capillaries in the muscles and more blood flow through, thus removing the products. Once the level of these products returns to normal, the sphincter contracts again.

Capillaries can also respond to changes in body temperature. As the working skeletal muscle generates heat the capillaries dilate to permit the "heated" blood to pass more easily towards the body surface.

Changes in blood distribution: adrenaline and the sympathetic nervous system act upon smooth muscle fibres in the walls of arteries. As a result blood vessels in the skin and abdominal organs undergo vasoconstriction so that blood which is normally "stored" in these organs is put into more active circulation. At the same time vasodilation permits a greater blood flow through the coronary vessels and the skeletal muscles. There is not enough blood in the circulation to fill the whole of the circulatory system in the dilated state.

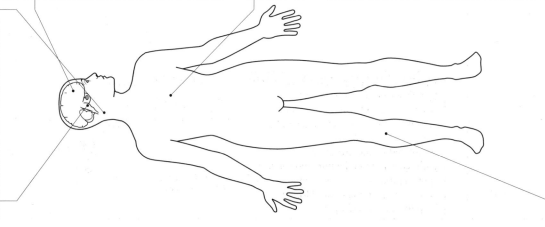

arteriole/capillary sphincter

shunt

capillary

venule

arteriole

pre-capillary sphincter

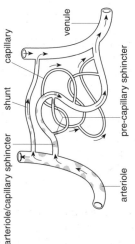

Natural defence systems of the body

prevent infection and disease.

PATHOGENS CAUSE DISEASE.

Many organisms may colonize the body of a human - the body is **warm**, **moist** and a **good food source**. These organisms may compete with human cells for nutrients or may produce by-products which are poisonous to human cells. This will affect the normal function of the cells, i.e. it will cause 'disease'. There are three major groups of pathogens:

Type of pathogen	Disease	Symptoms
VIRUS Protein coat Nucleic acid	Influenza	Fever (raised body temperature). Aching joints. Breathing problems. **Control by pain relief/ rest/drinking fluids**.
BACTERIUM Slime coat Cell wall 'Naked' DNA (not in chromosome)	Gonorrhoea	Painful to urinate - yellow discharge from penis or vagina. In the long-term may cause blockage of sperm ducts/oviducts leading to sterility. **Control with antibiotics**.
FUNGUS DNA in nucleus Hypha secretes enzymes into food	Athlete's foot	Irritation to moist areas of skin (e.g. between toes). 'Cracked' skin may become infected. **Control with fungicide/drying powders**.

PHAGOCYTES DESTROY PATHOGENS BY INGESTING THEM

Phagocytes are large white blood cells. They are 'attracted to' wounds or sites of infection by chemical messages. They leave the blood vessels and destroy any pathogens which they recognize.

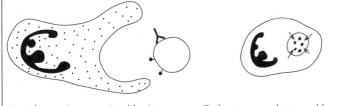

A pathogen is recognized by its surface proteins or by antibodies which 'label' it as dangerous.

Pathogens are destroyed by digestive enzymes secreted into 'food sac'.

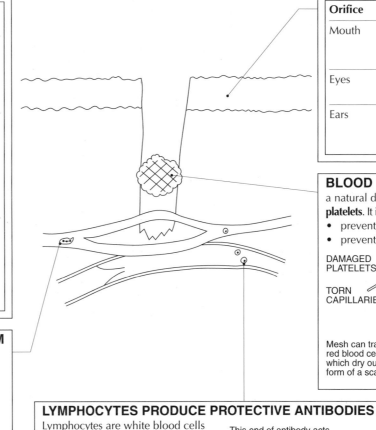

SKIN IS THE FIRST BARRIER:

the outer layer of the skin, the **epidermis**, is waxy and impermeable to water and to pathogens (although many micro-organisms can live on its surface). Where there are natural 'gaps' in the skin there may be protective secretions to prevent entry of pathogens, e.g.

Orifice	Function	Protected by
Mouth	Entry of food	Hydrochloric acid in stomach
Eyes	Entry of light	Lysozyme in tears
Ears	Entry of sound	Bacteriocidal ('bacteria-killing') wax

BLOOD CLOTTING 'PLUGS' WOUNDS

is a natural defence, largely due to **blood proteins** and **platelets**. It is able to block any unnatural gaps in the skin.

- prevents excessive blood loss
- prevents entry of pathogens

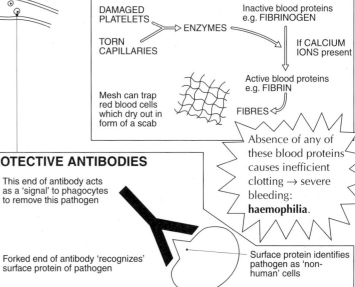

DAMAGED PLATELETS
TORN CAPILLARIES
→ ENZYMES →
Inactive blood proteins e.g. FIBRINOGEN
If CALCIUM IONS present
Active blood proteins e.g. FIBRIN
FIBRES

Mesh can trap red blood cells which dry out in form of a scab

Absence of any of these blood proteins causes inefficient clotting → severe bleeding: **haemophilia**.

LYMPHOCYTES PRODUCE PROTECTIVE ANTIBODIES

Lymphocytes are white blood cells which are found both in the blood and in lymph nodes (swellings in the lymphatic system). They are stimulated by the presence of pathogens to manufacture and release special proteins called **antibodies** which can recognize, bind to, and help to destroy pathogens.

This end of antibody acts as a 'signal' to phagocytes to remove this pathogen

Forked end of antibody 'recognizes' surface protein of pathogen

Surface protein identifies pathogen as 'non-human' cells

Inflammation and phagocytosis are part of the non-specific defence reactions of the body.

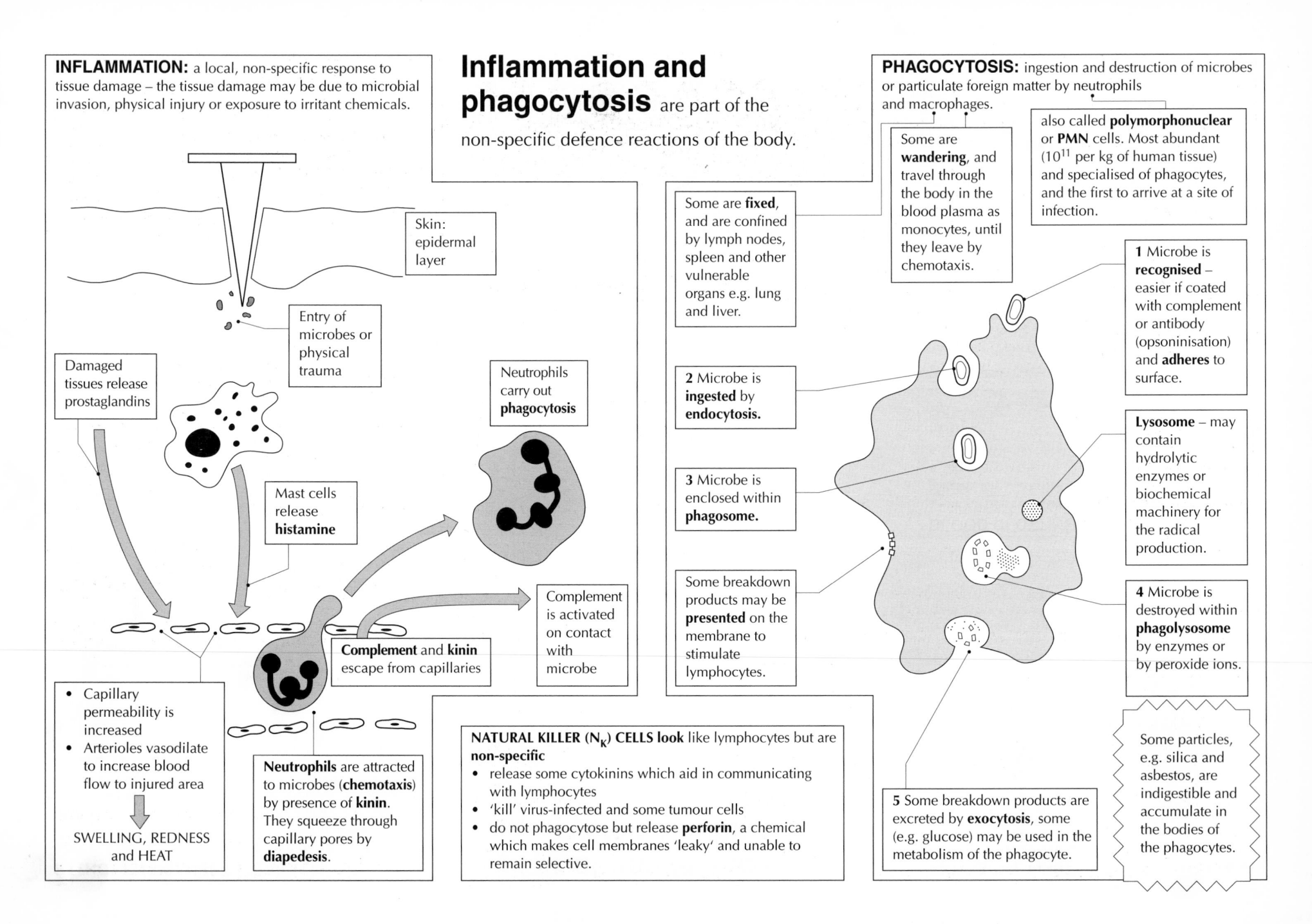

INFLAMMATION: a local, non-specific response to tissue damage – the tissue damage may be due to microbial invasion, physical injury or exposure to irritant chemicals.

Skin: epidermal layer

Entry of microbes or physical trauma

Damaged tissues release prostaglandins

Mast cells release **histamine**

Neutrophils carry out **phagocytosis**

Complement and **kinin** escape from capillaries

Complement is activated on contact with microbe

- Capillary permeability is increased
- Arterioles vasodilate to increase blood flow to injured area

SWELLING, REDNESS and HEAT

Neutrophils are attracted to microbes (**chemotaxis**) by presence of **kinin**. They squeeze through capillary pores by **diapedesis**.

NATURAL KILLER (N$_K$) CELLS look like lymphocytes but are **non-specific**
- release some cytokinins which aid in communicating with lymphocytes
- 'kill' virus-infected and some tumour cells
- do not phagocytose but release **perforin**, a chemical which makes cell membranes 'leaky' and unable to remain selective.

PHAGOCYTOSIS: ingestion and destruction of microbes or particulate foreign matter by neutrophils and macrophages.

Some are **fixed**, and are confined by lymph nodes, spleen and other vulnerable organs e.g. lung and liver.

Some are **wandering**, and travel through the body in the blood plasma as monocytes, until they leave by chemotaxis.

also called **polymorphonuclear** or **PMN** cells. Most abundant (10^{11} per kg of human tissue) and specialised of phagocytes, and the first to arrive at a site of infection.

1 Microbe is **recognised** – easier if coated with complement or antibody (opsoninisation) and **adheres** to surface.

2 Microbe is **ingested** by **endocytosis.**

3 Microbe is enclosed within **phagosome.**

Some breakdown products may be **presented** on the membrane to stimulate lymphocytes.

Lysosome – may contain hydrolytic enzymes or biochemical machinery for the radical production.

4 Microbe is destroyed within **phagolysosome** by enzymes or by peroxide ions.

5 Some breakdown products are excreted by **exocytosis**, some (e.g. glucose) may be used in the metabolism of the phagocyte.

Some particles, e.g. silica and asbestos, are indigestible and accumulate in the bodies of the phagocytes.

The immune response

involves a wide range of cells and their products in defence against diseases.

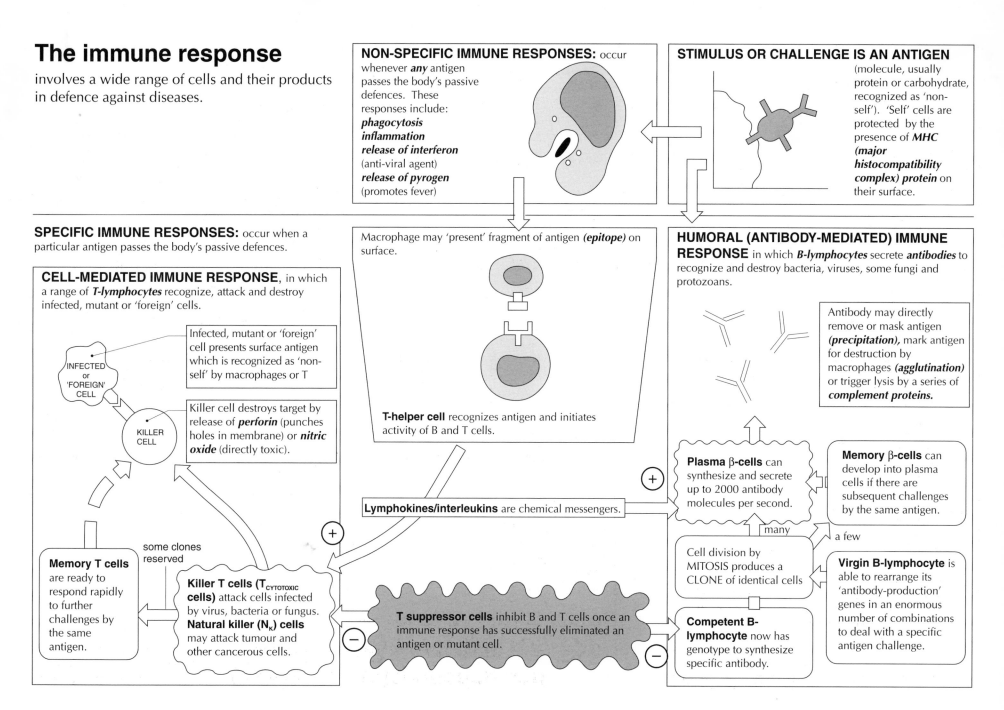

NON-SPECIFIC IMMUNE RESPONSES: occur whenever **any** antigen passes the body's passive defences. These responses include:
phagocytosis
inflammation
release of interferon (anti-viral agent)
release of pyrogen (promotes fever)

STIMULUS OR CHALLENGE IS AN ANTIGEN (molecule, usually protein or carbohydrate, recognized as 'non-self'). 'Self' cells are protected by the presence of **MHC (major histocompatibility complex) protein** on their surface.

SPECIFIC IMMUNE RESPONSES: occur when a particular antigen passes the body's passive defences.

CELL-MEDIATED IMMUNE RESPONSE, in which a range of **T-lymphocytes** recognize, attack and destroy infected, mutant or 'foreign' cells.

INFECTED or 'FOREIGN' CELL

Infected, mutant or 'foreign' cell presents surface antigen which is recognized as 'non-self' by macrophages or T

KILLER CELL

Killer cell destroys target by release of **perforin** (punches holes in membrane) or **nitric oxide** (directly toxic).

Macrophage may 'present' fragment of antigen **(epitope)** on surface.

T-helper cell recognizes antigen and initiates activity of B and T cells.

HUMORAL (ANTIBODY-MEDIATED) IMMUNE RESPONSE in which **B-lymphocytes** secrete **antibodies** to recognize and destroy bacteria, viruses, some fungi and protozoans.

Antibody may directly remove or mask antigen **(precipitation),** mark antigen for destruction by macrophages **(agglutination)** or trigger lysis by a series of **complement proteins.**

Lymphokines/interleukins are chemical messengers.

Plasma β-cells can synthesize and secrete up to 2000 antibody molecules per second.

Memory β-cells can develop into plasma cells if there are subsequent challenges by the same antigen.

many

a few

Memory T cells are ready to respond rapidly to further challenges by the same antigen.

some clones reserved

Killer T cells (T_CYTOTOXIC cells) attack cells infected by virus, bacteria or fungus.
Natural killer (N_K) cells may attack tumour and other cancerous cells.

T suppressor cells inhibit B and T cells once an immune response has successfully eliminated an antigen or mutant cell.

Cell division by MITOSIS produces a CLONE of identical cells

Competent B-lymphocyte now has genotype to synthesize specific antibody.

Virgin B-lymphocyte is able to rearrange its 'antibody-production' genes in an enormous number of combinations to deal with a specific antigen challenge.

Immune response II: antibodies, immunity and vaccination

IMMUNITY MAY BE ENHANCED

In **active immunity** an individual is provoked to **manufacture his or her own antibodies**.

a. **natural**

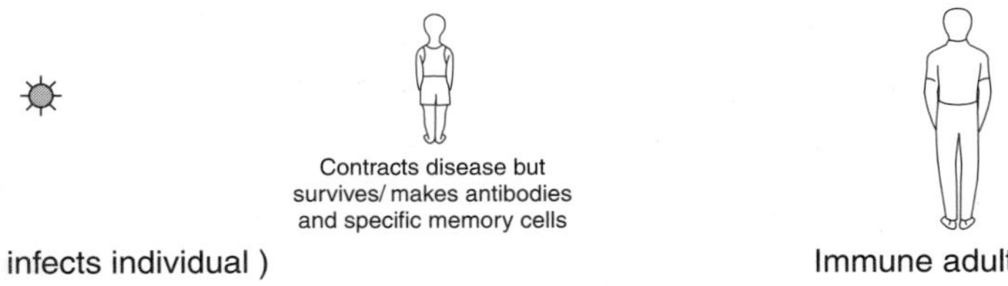

(Pathogen infects individual)

Contracts disease but survives/makes antibodies and specific memory cells

Immune adult

b. **artificial**

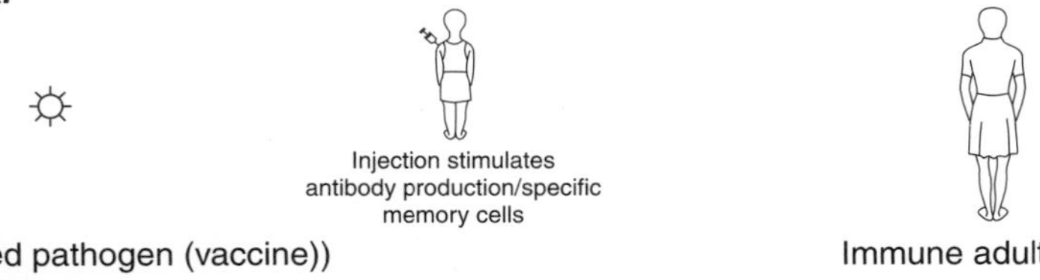

(Weakened pathogen (vaccine))

Injection stimulates antibody production/specific memory cells

Immune adult

In **passive immunity** an individual is protected by a **supply of pre-formed antibodies**.

a. **natural**

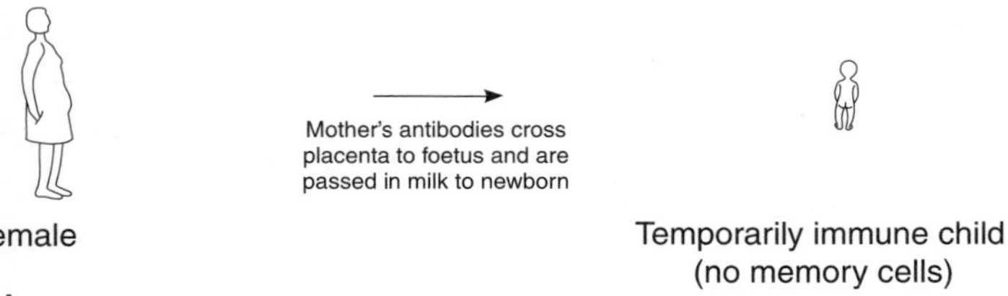

Immune female

Mother's antibodies cross placenta to foetus and are passed in milk to newborn

Temporarily immune child (no memory cells)

b. **artificial**

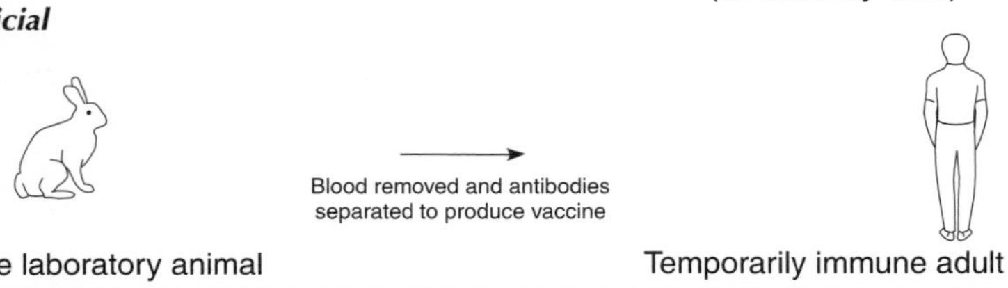

Immune laboratory animal

Blood removed and antibodies separated to produce vaccine

Temporarily immune adult

IMMUNISATION/ VACCINATION

provokes production of B and T **memory cells** which confer protection against a second challenge from the same antigen.

Secondary response: curve is steeper, peak is higher (commonly 10^3 x the primary response), lag period is negligible **due to the presence of B-memory cells.** Dominant antibody is IgG which is more stable and has a greater affinity for the antigen.

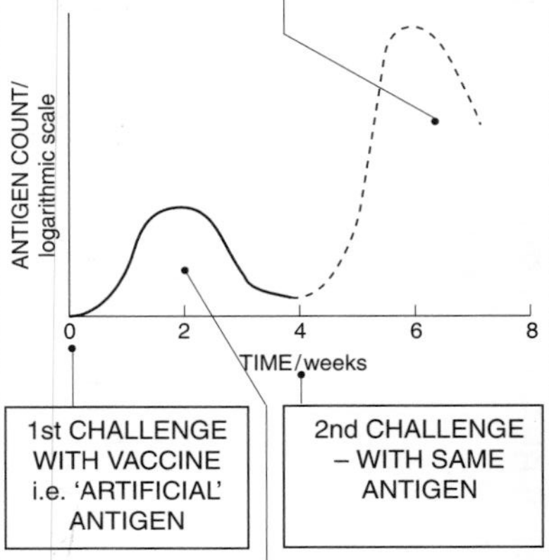

ANTIGEN COUNT/ logarithmic scale

TIME/weeks

1st CHALLENGE WITH VACCINE i.e. 'ARTIFICIAL' ANTIGEN

2nd CHALLENGE – WITH SAME ANTIGEN

Primary response: typical lag period is 3 days with peak at 11–14 days. Dominant antibody molecule is IgM.

LONG-TERM GOAL: THE ONE-SHOT VACCINE

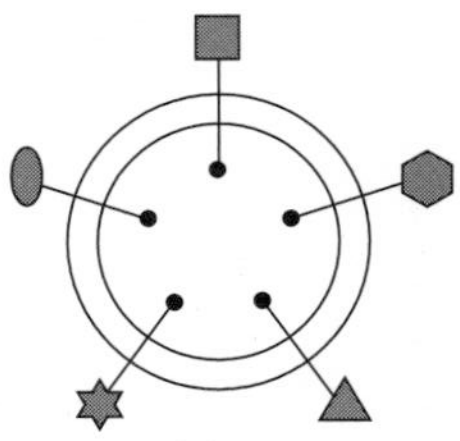

Use *Vaccinia* virus to 'carry' antigens to all common diseases e.g. measles, tetanus, diphtheria, polio and whooping cough.

VACCINES ARE OF THREE TYPES

	'KILLED' – inject dead mocribe	'LIVE' – inject weakened (attenuated) microbe	INACTIVATED TOXIN (TOXOIDS)
ADVANTAGE	Provokes immune response since surface antigens remain intact.	Low dosage since organisms multiply Offers longer term protection.	Good response. Safe to handle.
DISADVANTAGE	Several injections of high vaccine dose are required.	Microbe may mutate to original virulent form. Must be kept refrigerated.	Expensive production. Only applicable to microbes which produce toxin.
EXAMPLE	Salk polio vaccine.	Sabin (oral) polio vaccine.	Anti-diptheria and tetanus vaccines.

Problems associated with the immune response

Allergies occur when the immune system responds excessively to a common environmental challenge, such as house dust or pollen.

- may provoke sneezing, swollen and overactive mucous membranes, 'runny' eyes, itching and, in extreme cases, breathing difficulties
- 'allergic' individuals produce an excess of IgE antibodies as well as the more usual IgG type.
- binding of antigen to IgG on most cell surfaces causes release of **Histamine** and associated inflammation.

Hypersensitivity is an extreme and rapid allergic response. In the most severe cases **Anaphylaxis** may occur – this involves vasodilation and capillary leakage and the consequent lowering of blood pressure may even lead to death.

Transplant rejection occurs when T-lymphocytes recognise, invade and destroy transplanted 'foreign' tissues.

T_{HELPER} cells recognise 'foreign' protein on tissue and secrete interleukin.

Immature $T_{CYTOTOXIC}$ cells recognise antigen, and, under stimulation from IL–2, divide and mature into fully active $T_{CYTOTOXIC}$ cells.

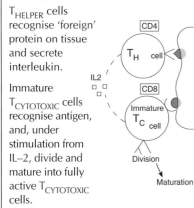

Mature $T_{CYTOTOXIC}$ cells invade the tissue and destroy cells carrying the 'foreign' protein: tissue destruction causes loss of function of transplanted organ.

Precautions to minimise risk of rejection
- close tissue matching between donor and recipient
- irradiation of bone marrow and lymph tissue – reduces lymphocyte production and thus reduces rejection
- immunosuppression – cyclosporin B reduces T_c numbers but B lymphocytes unaffected. Thus humoral immunity is retained.

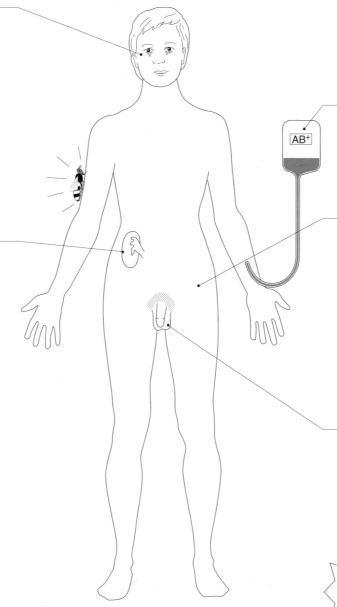

Blood transfusion may cause **agglutination** of red cells. Agglutinated blood may block capillaries, causing depletion of oxygen and glucose to respiring tissues. Agglutination is prevented by matching blood types between donor and recipient so that only **compatible** transfusions take place.

Autoimmunity arises when the 'immune' system fails to distinguish between 'self' and 'non-self' (foreign) cells, and attacks 'self', or body tissues.
- may affect one organ ('organ-specific' e.g. pernicious anaemia) or many ('organ non-specific' or 'multi-systemic' e.g. mixed connective tissue disease)
- causes generally not known, but may involve genetic, viral and environmental factors
- may be implicated in many conditions e.g. Rheumatoid arthritis
  Type I (Insulin-dependent) diabetes mellitus
  Multiple sclerosis
  Haemolytic anaemia

'Immune camouflage' is necessary for
- Sperm – hapoid so 'non-self' and head is 'hidden' within Sertoli cells of the seminiferous tubules.
- Foetus – carries some paternal antigens so must be protected from maternal immune surveillance by the placenta.

Vaccination/immunisation deliberately provokes the immune system. Incorrect dosage or extreme sensitivity to antigen may lead to an allergic response

Monoclonal antibodies: production and applications

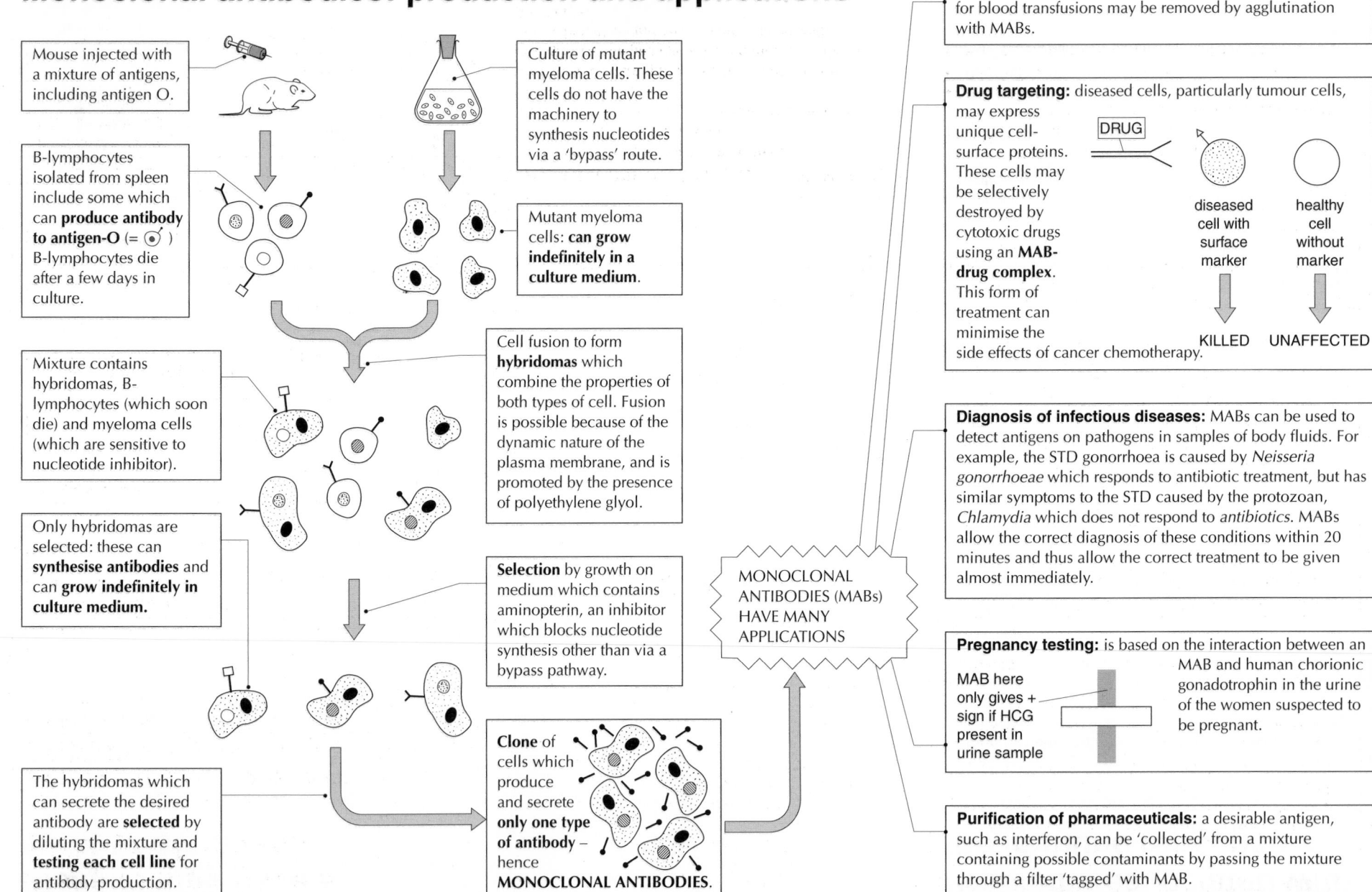

Mouse injected with a mixture of antigens, including antigen O.

Culture of mutant myeloma cells. These cells do not have the machinery to synthesis nucleotides via a 'bypass' route.

B-lymphocytes isolated from spleen include some which can **produce antibody to antigen-O** (= ⊙) B-lymphocytes die after a few days in culture.

Mutant myeloma cells: **can grow indefinitely in a culture medium.**

Mixture contains hybridomas, B-lymphocytes (which soon die) and myeloma cells (which are sensitive to nucleotide inhibitor).

Cell fusion to form **hybridomas** which combine the properties of both types of cell. Fusion is possible because of the dynamic nature of the plasma membrane, and is promoted by the presence of polyethylene glyol.

Only hybridomas are selected: these can **synthesise antibodies** and can **grow indefinitely in culture medium.**

Selection by growth on medium which contains aminopterin, an inhibitor which blocks nucleotide synthesis other than via a bypass pathway.

The hybridomas which can secrete the desired antibody are **selected** by diluting the mixture and **testing each cell line** for antibody production.

Clone of cells which produce and secrete **only one type of antibody** – hence **MONOCLONAL ANTIBODIES.**

MONOCLONAL ANTIBODIES (MABs) HAVE MANY APPLICATIONS

Blood screening: undesirable proteins in blood to be used for blood transfusions may be removed by agglutination with MABs.

Drug targeting: diseased cells, particularly tumour cells, may express unique cell-surface proteins. These cells may be selectively destroyed by cytotoxic drugs using an **MAB-drug complex**. This form of treatment can minimise the side effects of cancer chemotherapy.

DRUG

diseased cell with surface marker

healthy cell without marker

KILLED   UNAFFECTED

Diagnosis of infectious diseases: MABs can be used to detect antigens on pathogens in samples of body fluids. For example, the STD gonorrhoea is caused by *Neisseria gonorrhoeae* which responds to antibiotic treatment, but has similar symptoms to the STD caused by the protozoan, *Chlamydia* which does not respond to *antibiotics*. MABs allow the correct diagnosis of these conditions within 20 minutes and thus allow the correct treatment to be given almost immediately.

Pregnancy testing: is based on the interaction between an MAB and human chorionic gonadotrophin in the urine of the women suspected to be pregnant.

MAB here only gives + sign if HCG present in urine sample

Purification of pharmaceuticals: a desirable antigen, such as interferon, can be 'collected' from a mixture containing possible contaminants by passing the mixture through a filter 'tagged' with MAB.

Disease may have a number of causes

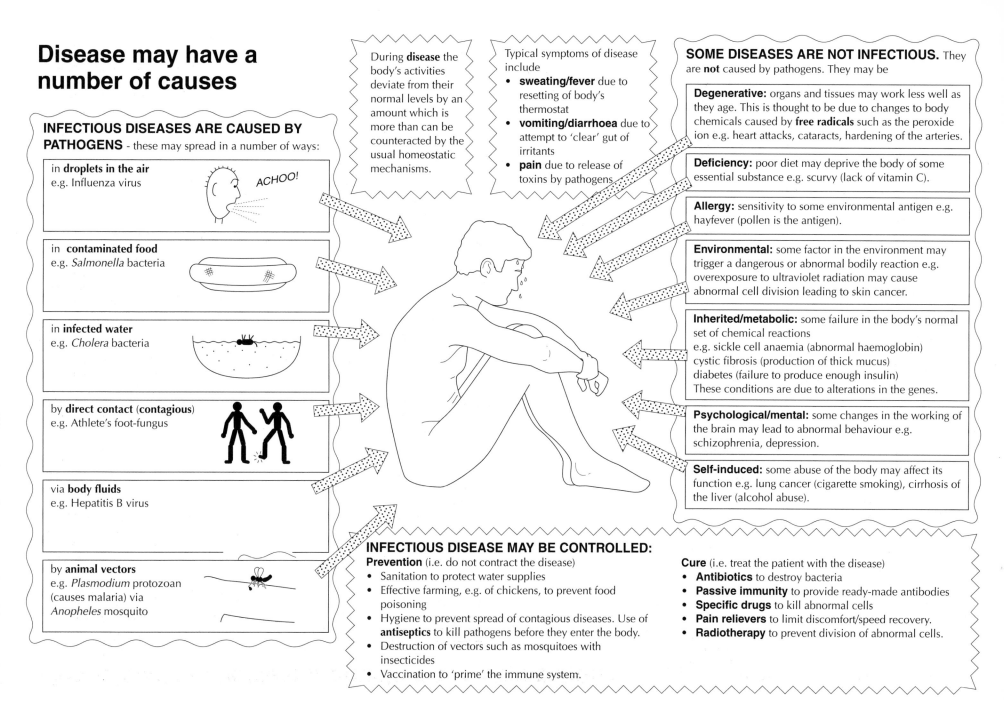

INFECTIOUS DISEASES ARE CAUSED BY PATHOGENS - these may spread in a number of ways:

in droplets in the air
e.g. Influenza virus

ACHOO!

in contaminated food
e.g. *Salmonella* bacteria

in infected water
e.g. *Cholera* bacteria

by direct contact (contagious)
e.g. Athlete's foot-fungus

via body fluids
e.g. Hepatitis B virus

by animal vectors
e.g. *Plasmodium* protozoan (causes malaria) via *Anopheles* mosquito

During **disease** the body's activities deviate from their normal levels by an amount which is more than can be counteracted by the usual homeostatic mechanisms.

Typical symptoms of disease include
- **sweating/fever** due to resetting of body's thermostat
- **vomiting/diarrhoea** due to attempt to 'clear' gut of irritants
- **pain** due to release of toxins by pathogens

SOME DISEASES ARE NOT INFECTIOUS. They are **not** caused by pathogens. They may be

Degenerative: organs and tissues may work less well as they age. This is thought to be due to changes to body chemicals caused by **free radicals** such as the peroxide ion e.g. heart attacks, cataracts, hardening of the arteries.

Deficiency: poor diet may deprive the body of some essential substance e.g. scurvy (lack of vitamin C).

Allergy: sensitivity to some environmental antigen e.g. hayfever (pollen is the antigen).

Environmental: some factor in the environment may trigger a dangerous or abnormal bodily reaction e.g. overexposure to ultraviolet radiation may cause abnormal cell division leading to skin cancer.

Inherited/metabolic: some failure in the body's normal set of chemical reactions
e.g. sickle cell anaemia (abnormal haemoglobin)
cystic fibrosis (production of thick mucus)
diabetes (failure to produce enough insulin)
These conditions are due to alterations in the genes.

Psychological/mental: some changes in the working of the brain may lead to abnormal behaviour e.g. schizophrenia, depression.

Self-induced: some abuse of the body may affect its function e.g. lung cancer (cigarette smoking), cirrhosis of the liver (alcohol abuse).

INFECTIOUS DISEASE MAY BE CONTROLLED:

Prevention (i.e. do not contract the disease)
- Sanitation to protect water supplies
- Effective farming, e.g. of chickens, to prevent food poisoning
- Hygiene to prevent spread of contagious diseases. Use of **antiseptics** to kill pathogens before they enter the body.
- Destruction of vectors such as mosquitoes with insecticides
- Vaccination to 'prime' the immune system.

Cure (i.e. treat the patient with the disease)
- **Antibiotics** to destroy bacteria
- **Passive immunity** to provide ready-made antibodies
- **Specific drugs** to kill abnormal cells
- **Pain relievers** to limit discomfort/speed recovery.
- **Radiotherapy** to prevent division of abnormal cells.

Human immunodeficiency virus (HIV)

may depress the immune response

Protease enzymes cleave large proteins into subunits for viral coat assembly

Acquired Immune Deficiency Syndrome (AIDS) is caused by a depression of the immune response

may be due to HIV infection of dendritic cells (display antigen) or T$_{HELPER}$ lymphocytes (coordinate lymphocyte action).

allows attack by normally harmless organisms, e.g.
Pneumocystis carinii → pneumonia
Candida albicans → thrush
Cytomegalovirus → retinal disease
and characteristic skin lesions of Kaposi's sarcoma.

Glycoprotein 'spikes' pierce the outer lipid bilayer. These spikes enable the virus to bind to receptors (CD4 receptors) on the surface membrane of susceptible cells. Such cells include T$_{HELPER}$ lymphocytes, macrophages, microglial cells in the brain and dendritic cells. Other cells may be infected, at least in the laboratory, which means that CD4 may not be the only receptor.

Viral coat assembly may be prevented by protease inhibitors

RNA – there are two strands, as in most retroviruses.

Protein coats enclose and support the viral RNA and enzymes.

Envelope membrane is a lipid bilayer which
a. anchors the gp 41 glycoporotein
b. fuses with the host cell membrane to allow the HIV to enter its target.

HIV enters cell – viral RNA is transcribed to viral DNA which is incorporated into host DNA.

New HIV particles are assembled from proteins made on instruction of viral DNA.

Blocking entry – one possibility for prevention is to prevent gp120–CD4 binding.

Soluble CD4 (immunoadhesin) binds to viral particles

Anti-CD4 binds onto CD4 molecules

Integrase is the enzyme needed to 'cut and paste' the viral DNA into the host DNA. Other enzyme functions include combination of separate 'genes' into a complete viral genome.

One potential drug treatment involves an enzyme inhibitor which prevents assembly of the viral genome.

Reverse transcriptase is the typical retroviral enzyme. It can transcribe the viral RNA molecule into a section of DNA which can then be integrated into the host DNA. In this form the DNA is protected from the host's immune system, although the host can produce antibodies to some of the viral proteins. This DNA contains

| gag | pol | cnv | reg |

codes for inner coat proteins

codes for reverse transcriptase and integrase

codes for gp41 and gp120 glycoproteins

codes for regulatory genes which allow the integrated virus to exist in three states of activity – inactive, slow or rapid.

Inhibition of viral replication has been the most widespread strategy for prevention of infection e.g. **zidovudine** (AZT) is an inhibitor of reverse transcriptase.

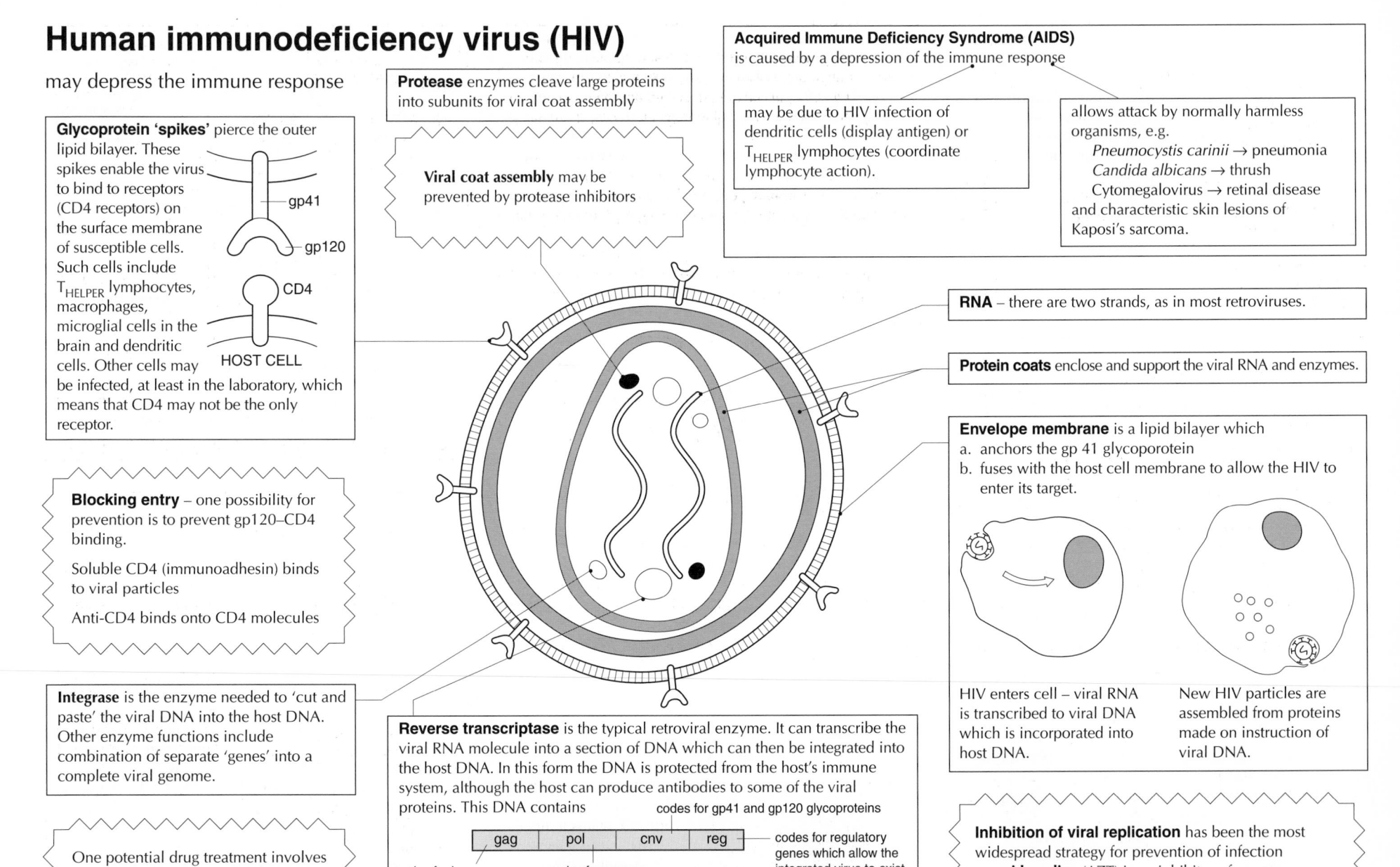

gp41
gp120
CD4
HOST CELL

PENICILLIUM MOULD is a **saprotroph** which reproduces by **sporulation**.

Other, related topics can be found on pages ....

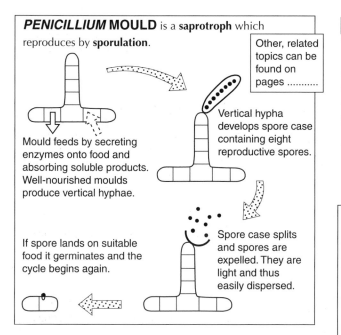

Mould feeds by secreting enzymes onto food and absorbing soluble products. Well-nourished moulds produce vertical hyphae.

Vertical hypha develops spore case containing eight reproductive spores.

If spore lands on suitable food it germinates and the cycle begins again.

Spore case splits and spores are expelled. They are light and thus easily dispersed.

LARGE SCALE PRODUCTION OF PENICILLIN

is a good example of the commercial use of fermentation reactions.

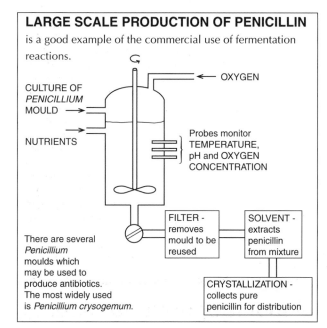

CULTURE OF *PENICILLIUM* MOULD

NUTRIENTS

OXYGEN

Probes monitor TEMPERATURE, pH and OXYGEN CONCENTRATION

There are several *Penicillium* moulds which may be used to produce antibiotics. The most widely used is *Penicillium crysogemum*.

FILTER - removes mould to be reused

SOLVENT - extracts penicillin from mixture

CRYSTALLIZATION - collects pure penicillin for distribution

Penicillin is an antibiotic

The *Penicillium* mould may make products which it secretes into its environment to kill off any disease-causing or competitive micro-organisms. **A compound made by a mould to kill off any other micro-organism is called an ANTIBIOTIC.**

Antibiotics are valuable in the control of bacterial diseases because they affect bacterial cells but rarely affect human or animal cells.

ACTION OF PENICILLIN

Bacteria multiply very rapidly by **binary fission** i.e. they grow and then divide into two. This can happen very rapidly (every 20 minutes), causing disease as the bacterial metabolism competes with or inhibits the activity of human cells.

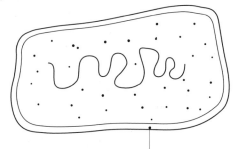

Bacterial cells are surrounded by a **cell wall**. **Penicillin prevents the bacterium from making components of the cell wall** - the bacterial cell is therefore weakened.

Antibiotics may be **bacteriocidal** (i.e. they kill the pathogenic bacterium directly) or they may be **bacteriostatic** (i.e. they prevent replication, leaving the host's defences to kill the existing pathogens).

Penicillin is bacteriocidal at high concentrations, but this may have some side-effects in humans and encourage development of strains of bacteria resistant to treatment by penicillin. **Penicillin is usually given at doses which are bacteriostatic.**

Some important diseases caused by bacteria and thus suitable for antibiotic treatment are:

Dysentery
Food poisoning } affect gut

Syphilis
Gonorrhoea } affect reproductive organs

Pneumonia - affects lungs

Botulism - affects nervous system

RESISTANT BACTERIAL STRAINS MAY DEVELOP

Bacteria breed rapidly and populations are enormous. Within these populations rare **mutations** may produce cells which are **antibiotic-resistant**. Over-use of antibiotics may destroy 'normal' bacterial cells but allow 'resistant' cells to survive. These may multiply to form a resistant poulation.

ANTIBIOTIC TREATMENT

CELL DIVISION

One resistant cell in population

Only resistant cell survives

Whole population is now resistant

This is a form of **artificial selection**.

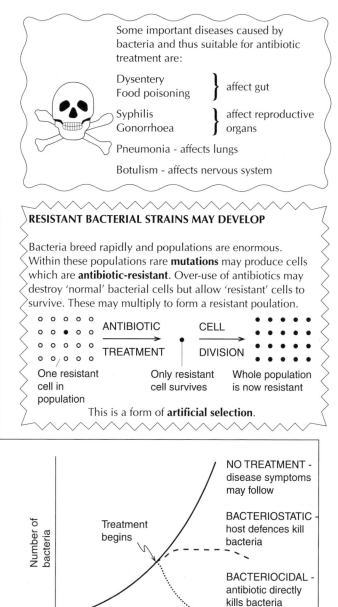

NO TREATMENT - disease symptoms may follow

BACTERIOSTATIC - host defences kill bacteria

BACTERIOCIDAL - antibiotic directly kills bacteria

Treatment begins

Number of bacteria

Time/h

Infection of host

Temperature control in endothermic organisms

is an example of **negative feedback**. Any deviation from optimum conditions sets mechanisms in motion to cancel out the change.

The **temperature control centre** is located in the **hypothalamus** in the floor of the brain. It receives incoming messages about the body temperature and makes certain that the appropriate corrective mechanisms operate to cancel out any deviations.

Skin increases heat loss

Sweat is secreted and **evaporation** consumes heat from body.

Vasodilation of surface capillaries allows blood to **radiate** heat away from body.

Relaxation of hair erector muscles allows hair to lie flat against skin to permit maximum heat loss by convection.

RADIATING HEAT

The **optimum temperature** for the body's activities is about 37°C. This ensures that

- enzymes work rapidly but are not denatured
- cell membranes keep their structure
- blood does not become too viscous to pass through narrow capillaries.

DETECTOR → INTEGRATOR

INCREASE

NORM

CORRECTIVE MECHANISMS: attempt to increase heat loss

NORM

During **fever**, the body's temperature may be reset 2-3°C higher: the higher temperatures denature proteins in the infecting bacteria more rapidly than they affect human proteins.

Deviations from the norm are detected

a. by **hot** and **cold sensors** in the skin

b by sensors in the hypothalamus which measure blood temperature.

DECREASE

DETECTOR → INTEGRATOR

CORRECTIVE MECHANISMS: attempt to conserve heat

Hypothermia ("low temperature")

- occurs if body's core temperature falls below 35°C
- causes irrational behaviour, brain damage, circulatory problems
- elderly people are very susceptible because they may be immobile
- children are very susceptible because they have a high surface area to volume ratio.

Humans may also

shiver - muscular activity generates heat

add extra clothing - insulation reduces heat loss

eat more - eating stimulates heat production by respiration…

…or **turn on the central heating!**

(humans can control their **external** environment too!)

Contraction of hair erector muscles traps a layer of still air which reduces heat losses by **convection**.

Vasoconstriction of surface capillaries shunts blood away from skin and so reduces heat losses by **radiation**.

Sweat glands do not secrete sweat so there is no consumption of body heat in **evaporation**.

SKIN REDUCES HEAT LOSS

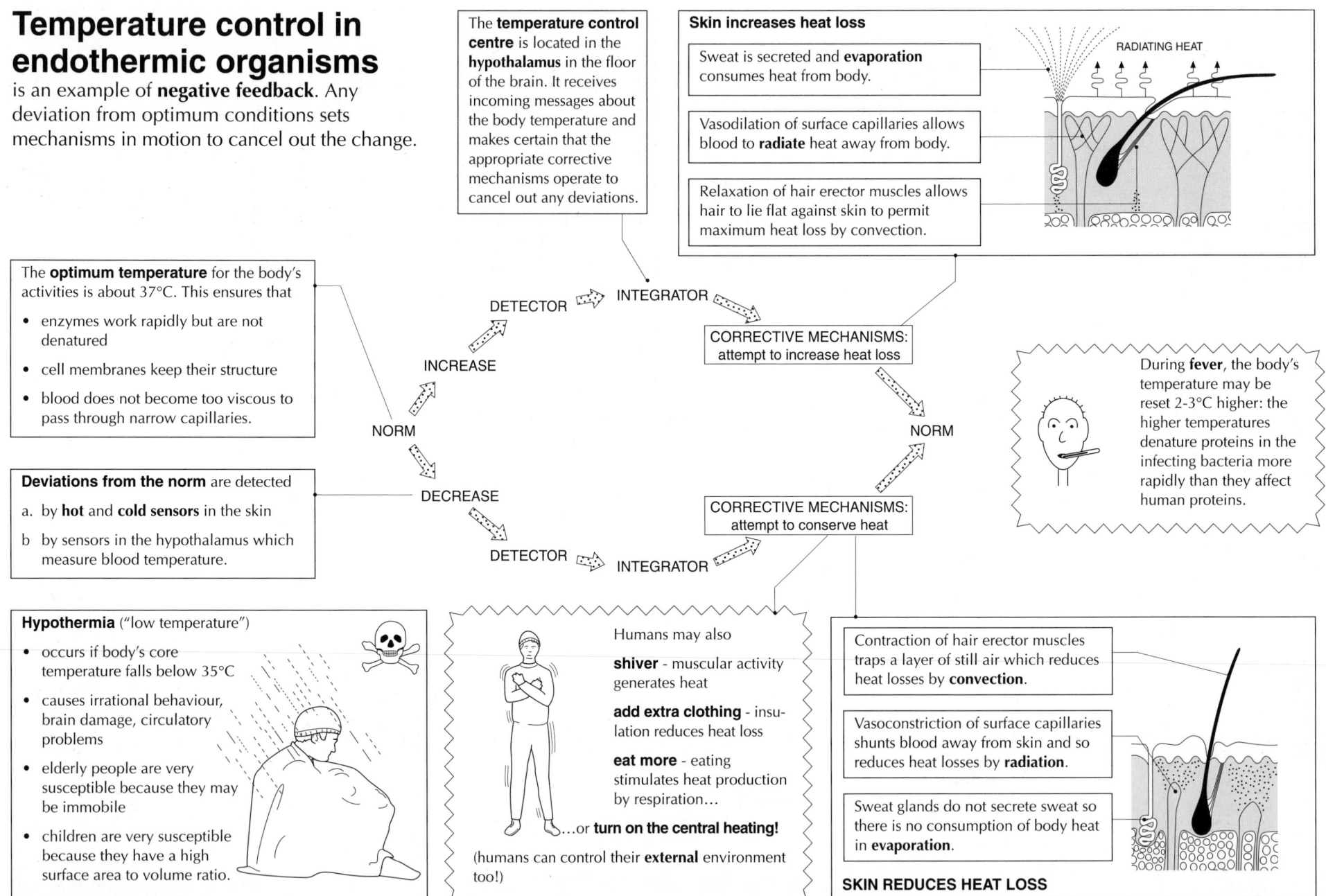

Control of body temperature in mammals

Changes in behaviour

Dressing/undressing; moving in and out of shade; rest/activity cycles may all affect heat loss or production.

Hypothalamus

contains the thermoregulatory centre which compares sensory input with a set point and initiates the appropriate motor responses. The set point may be raised by the action of pyrogens during pyrexia (fever).

Core temperature

affects temperature of circulating blood which is monitored in the thermoregulatory centre.

Skin temperature

is detected by skin thermoreceptors which deliver sensory input to hypothalamus via cutaneous nerves.

Tension/shivering in skeletal muscle. High energy cost and not effective for long periods.

EXTERNAL TEMPERATURE HIGH

Pilo-erector muscles relaxed: hair shafts 'flatten' and allow free circulation of air over hairs. Moving air is a good convector of heat.

EXTERNAL TEMPERATURE LOW

Pilo-erector muscles contracted: hair shafts perpendicular to skin surface. Trapped air is a poor conductor of heat so warm skin is insulated (similar effect by adding layers of clothing).

EXTERNAL TEMPERATURE HIGH

Sweat changed into vapour, taking latent heat of evaporation from the body to do this (about 2.5 kJ for each gram evaporated). Sweat glands extract larger volume of fluid from blood.

Fluid overflows here

Fluid flows up duct

EXTERNAL TEMPERATURE LOW

Skin surface comparatively dry – no evaporation and no cooling effect.

Sweat glands extract very little fluid from blood.

Non-shivering heat response

triggers a general increase in metabolic rate. This is particularly noticeable in **brown adipose tissue** of newborns and animals that become acclimatized to cold. The **liver** of adults is also affected.

Voluntary responses

Automatic responses

CEREBAL CORTEX

Releasing hormones

ANTERIOR PITUITARY GLAND

Thyrotrophic hormone

THYROID GLAND

THYROXINE

THYROXINE

ADRENALINE MEDULLA

ADRENALINE

Sympathetic neurones

Sympathetic neurones

VASOMOTOR CENTRE in/ medulla oblongata

EXTERNAL TEMPERATURE HIGH

Vasodilation: sphincters/dilation of superficial arterioles allow blood close to surface. Heat lost by radiation – body cooled.

heat

epidermis

dermis

EXTERNAL TEMPERATURE LOW

Vasoconstriction: superficial arterioles are constricted so that blood is shunted away from surface – heat is conserved.

epidermis

dermis

Summary:

THERMORECEPTORS
↓
INTEGRATION BY HYPOTHALAMUS
↓
APPROPRIATE RESPONSES

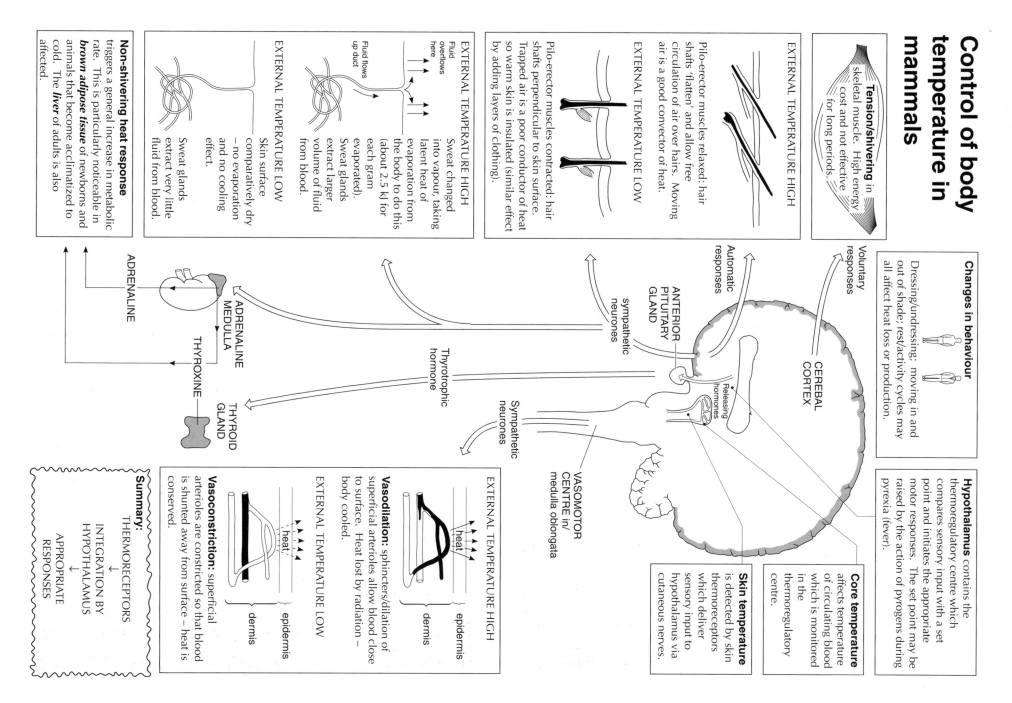

Blood glucose concentration

is regulated by negative feedback

Insulin a. increases uptake of glucose by cells of liver, muscle and adipose tissue
b. increases conversion of glucose to **glycogen** and **fats**
c. decreases hydrolysis of glycogen to glucose, and synthesis of glucose from amino acids.

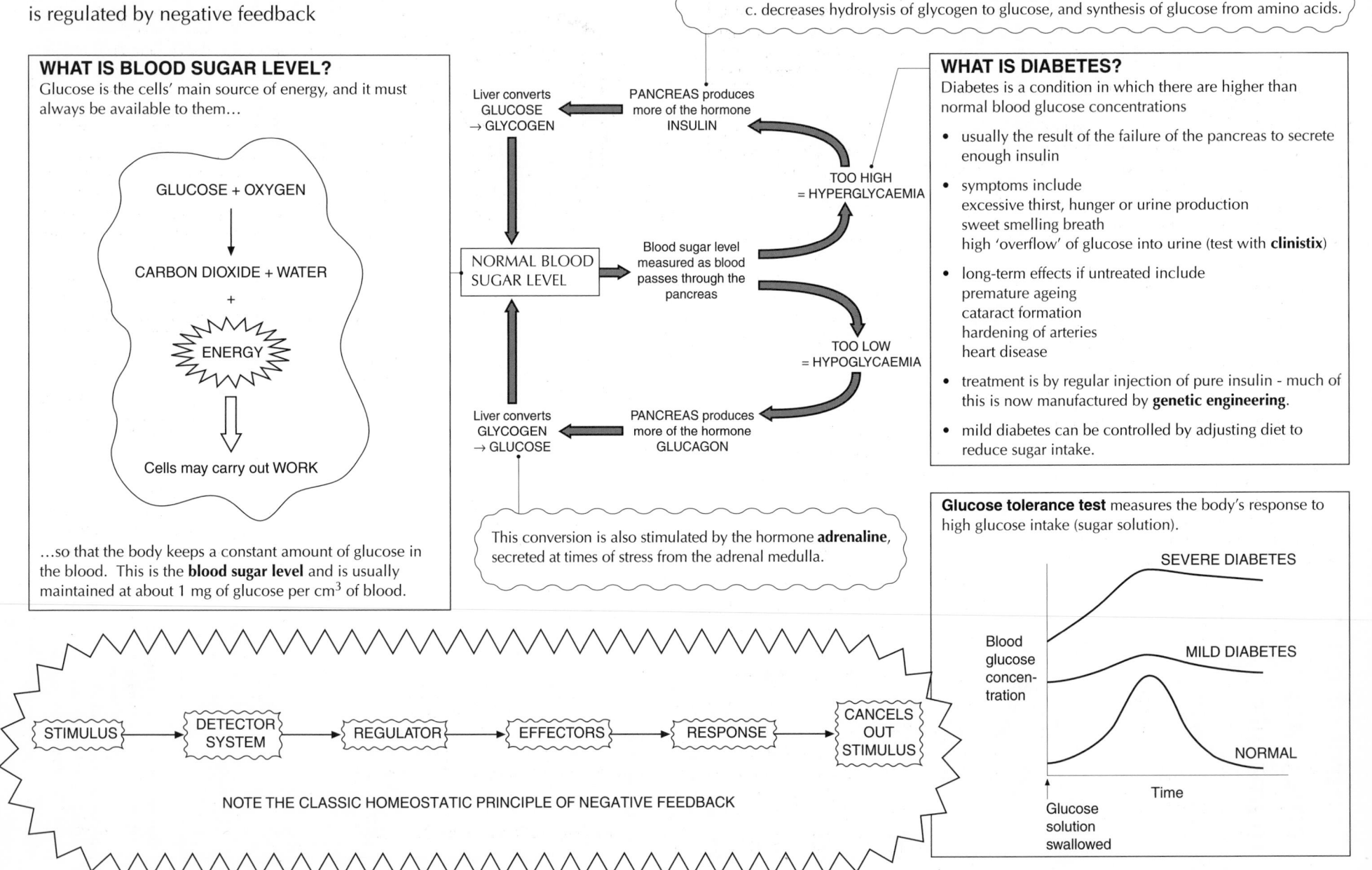

WHAT IS BLOOD SUGAR LEVEL?
Glucose is the cells' main source of energy, and it must always be available to them…

GLUCOSE + OXYGEN

↓

CARBON DIOXIDE + WATER

+

ENERGY

⇓

Cells may carry out WORK

…so that the body keeps a constant amount of glucose in the blood. This is the **blood sugar level** and is usually maintained at about 1 mg of glucose per cm^3 of blood.

Liver converts
GLUCOSE
→ GLYCOGEN

PANCREAS produces
more of the hormone
INSULIN

NORMAL BLOOD
SUGAR LEVEL

Blood sugar level
measured as blood
passes through the
pancreas

TOO HIGH
= HYPERGLYCAEMIA

TOO LOW
= HYPOGLYCAEMIA

Liver converts
GLYCOGEN
→ GLUCOSE

PANCREAS produces
more of the hormone
GLUCAGON

This conversion is also stimulated by the hormone **adrenaline**, secreted at times of stress from the adrenal medulla.

WHAT IS DIABETES?
Diabetes is a condition in which there are higher than normal blood glucose concentrations

- usually the result of the failure of the pancreas to secrete enough insulin

- symptoms include
  excessive thirst, hunger or urine production
  sweet smelling breath
  high 'overflow' of glucose into urine (test with **clinistix**)

- long-term effects if untreated include
  premature ageing
  cataract formation
  hardening of arteries
  heart disease

- treatment is by regular injection of pure insulin - much of this is now manufactured by **genetic engineering**.

- mild diabetes can be controlled by adjusting diet to reduce sugar intake.

Glucose tolerance test measures the body's response to high glucose intake (sugar solution).

SEVERE DIABETES

MILD DIABETES

Blood
glucose
concen-
tration

NORMAL

Time

Glucose
solution
swallowed

STIMULUS → DETECTOR SYSTEM → REGULATOR → EFFECTORS → RESPONSE → CANCELS OUT STIMULUS

NOTE THE CLASSIC HOMEOSTATIC PRINCIPLE OF NEGATIVE FEEDBACK

The liver is the largest gland,

weighing between 1.0 and 2.3 kg. It is situated in the upper abdomen, just beneath the diaphragm, and has four lobes. On the posterior surface is the **portal fissure** where various structures – hepatic portal vein, hepatic artery, hepatic vein, bile duct, lymph vessels – together with sympathetic and parasympathetic nerve fibres, enter or leave the gland.

BLOOD SUPPLY OF THE LIVER

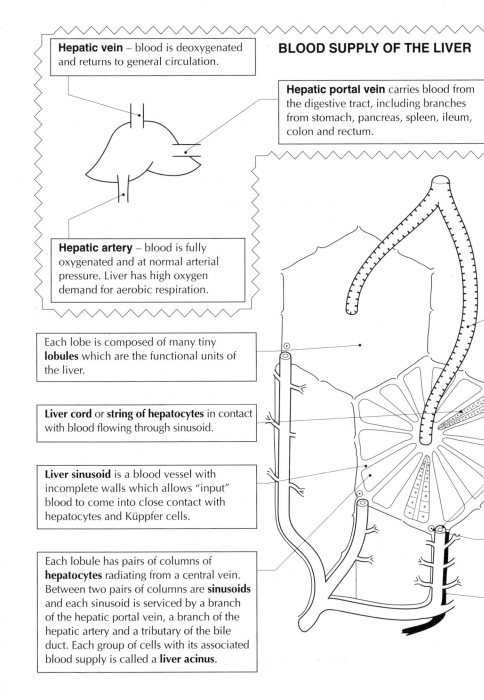

Hepatic vein – blood is deoxygenated and returns to general circulation.

Hepatic portal vein carries blood from the digestive tract, including branches from stomach, pancreas, spleen, ileum, colon and rectum.

Hepatic artery – blood is fully oxygenated and at normal arterial pressure. Liver has high oxygen demand for aerobic respiration.

Each lobe is composed of many tiny **lobules** which are the functional units of the liver.

Liver cord or **string of hepatocytes** in contact with blood flowing through sinusoid.

Liver sinusoid is a blood vessel with incomplete walls which allows "input" blood to come into close contact with hepatocytes and Küppfer cells.

Each lobule has pairs of columns of **hepatocytes** radiating from a central vein. Between two pairs of columns are **sinusoids** and each sinusoid is serviced by a branch of the hepatic portal vein, a branch of the hepatic artery and a tributary of the bile duct. Each group of cells with its associated blood supply is called a **liver acinus**.

Intralobular (central) vein: unites with others to form the **hepatic vein** which returns "modified" blood to the general circulation.

Interlobular vein is a branch of the portal vein and delivers blood, with varying solute concentration, from the gut.

Interlobular artery is a branch of the hypatic artery delivering oxygenated blood to the liver.

Bile canaliculus: unites with others to form the bile duct to take away the liver secretion, bile, from between the strings of hepatocytes.

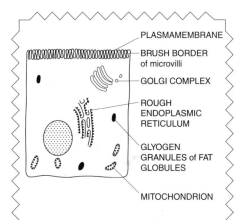

PLASMAMEMBRANE

BRUSH BORDER of microvilli

GOLGI COMPLEX

ROUGH ENDOPLASMIC RETICULUM

GLYOGEN GRANULES of FAT GLOBULES

MITOCHONDRION

The **hepatocyte** or liver cell has microvilli to increase surface area for absorption from the percolating blood in the sinusoids, prominent Golgi Complex for the preparation of cellular products for secretion, rough endoplasmic reticulum for the synthesis and intracellular transport of numerous proteins, glycogen granules or fat globules to act as energy reserves, numerous mitochondria to generate ATP for the liver's numerous metabolic reactions. The plasmamembrane has numerous transporter proteins as well as insulin receptors.

Although all hepatocytes perform all liver functions there is some localisation of function. For example, cells closest to the portal capillaries are most active in gluconeogenesis and glycogenesis and in oxidative ATP formation whilst those closest to the central vein are most active in fat synthesis, glycolysis and drug metabolism. It is these perivenous cells which are most prone to damage following paracetomol overdose.

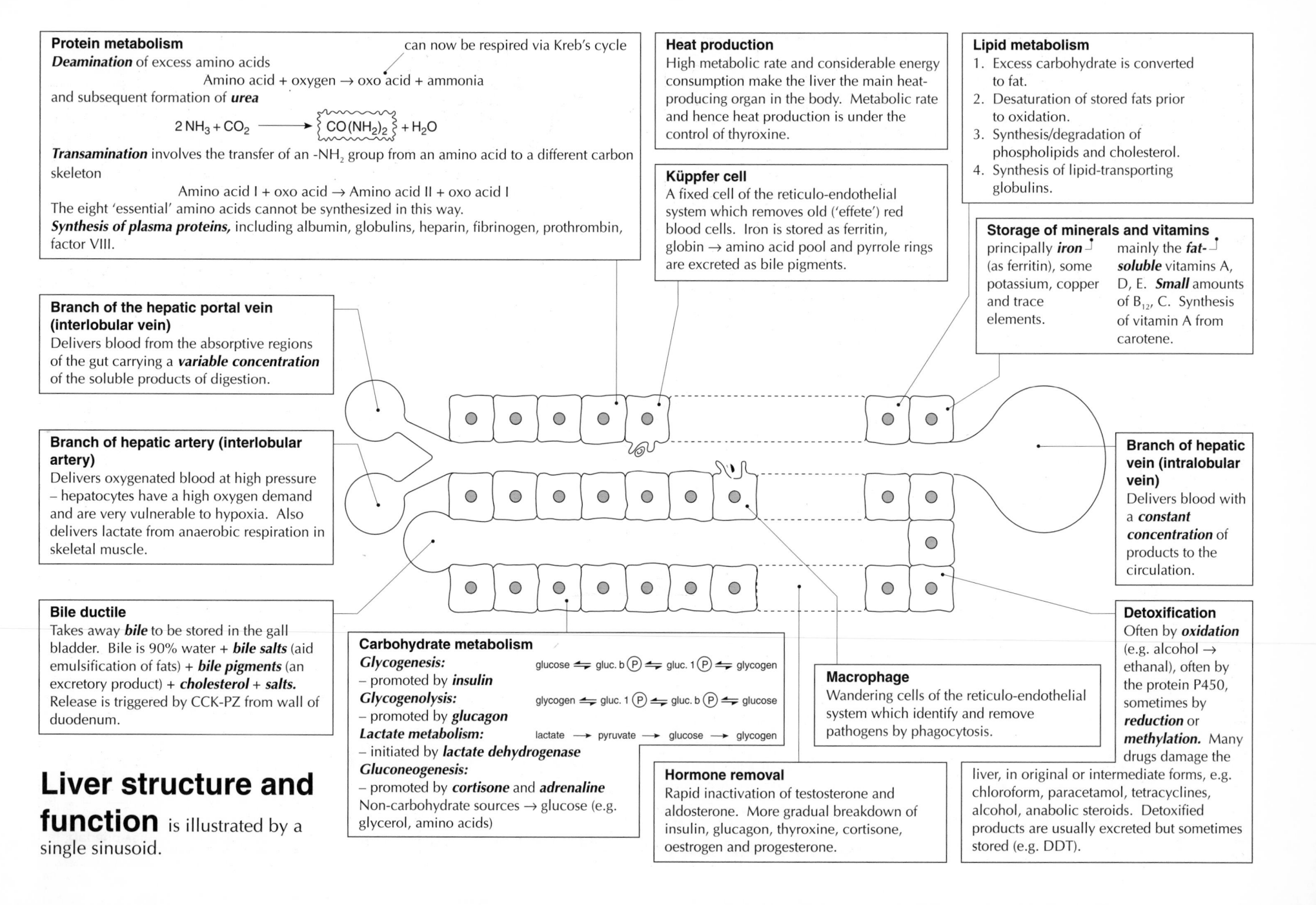

Protein metabolism
Deamination of excess amino acids
Amino acid + oxygen → oxo acid + ammonia [can now be respired via Kreb's cycle]
and subsequent formation of **urea**

$$2\,NH_3 + CO_2 \longrightarrow CO(NH_2)_2 + H_2O$$

Transamination involves the transfer of an -NH$_2$ group from an amino acid to a different carbon skeleton
Amino acid I + oxo acid → Amino acid II + oxo acid I
The eight 'essential' amino acids cannot be synthesized in this way.
Synthesis of plasma proteins, including albumin, globulins, heparin, fibrinogen, prothrombin, factor VIII.

Heat production
High metabolic rate and considerable energy consumption make the liver the main heat-producing organ in the body. Metabolic rate and hence heat production is under the control of thyroxine.

Küppfer cell
A fixed cell of the reticulo-endothelial system which removes old ('effete') red blood cells. Iron is stored as ferritin, globin → amino acid pool and pyrrole rings are excreted as bile pigments.

Lipid metabolism
1. Excess carbohydrate is converted to fat.
2. Desaturation of stored fats prior to oxidation.
3. Synthesis/degradation of phospholipids and cholesterol.
4. Synthesis of lipid-transporting globulins.

Storage of minerals and vitamins
principally **iron** (as ferritin), some potassium, copper and trace elements. mainly the **fat-soluble** vitamins A, D, E. **Small** amounts of B$_{12}$, C. Synthesis of vitamin A from carotene.

Branch of the hepatic portal vein (interlobular vein)
Delivers blood from the absorptive regions of the gut carrying a **variable concentration** of the soluble products of digestion.

Branch of hepatic artery (interlobular artery)
Delivers oxygenated blood at high pressure – hepatocytes have a high oxygen demand and are very vulnerable to hypoxia. Also delivers lactate from anaerobic respiration in skeletal muscle.

Branch of hepatic vein (intralobular vein)
Delivers blood with a **constant concentration** of products to the circulation.

Bile ductile
Takes away **bile** to be stored in the gall bladder. Bile is 90% water + **bile salts** (aid emulsification of fats) + **bile pigments** (an excretory product) + **cholesterol** + **salts.** Release is triggered by CCK-PZ from wall of duodenum.

Carbohydrate metabolism
Glycogenesis:
– promoted by **insulin**
glucose ⇌ gluc. b (P) ⇌ gluc. 1 (P) ⇌ glycogen
Glycogenolysis:
– promoted by **glucagon**
glycogen ⇌ gluc. 1 (P) ⇌ gluc. b (P) ⇌ glucose
Lactate metabolism:
– initiated by **lactate dehydrogenase**
lactate → pyruvate → glucose → glycogen
Gluconeogenesis:
– promoted by **cortisone** and **adrenaline**
Non-carbohydrate sources → glucose (e.g. glycerol, amino acids)

Macrophage
Wandering cells of the reticulo-endothelial system which identify and remove pathogens by phagocytosis.

Hormone removal
Rapid inactivation of testosterone and aldosterone. More gradual breakdown of insulin, glucagon, thyroxine, cortisone, oestrogen and progesterone.

Detoxification
Often by **oxidation** (e.g. alcohol → ethanal), often by the protein P450, sometimes by **reduction** or **methylation.** Many drugs damage the liver, in original or intermediate forms, e.g. chloroform, paracetamol, tetracyclines, alcohol, anabolic steroids. Detoxified products are usually excreted but sometimes stored (e.g. DDT).

Liver structure and function is illustrated by a single sinusoid.

The urinary system

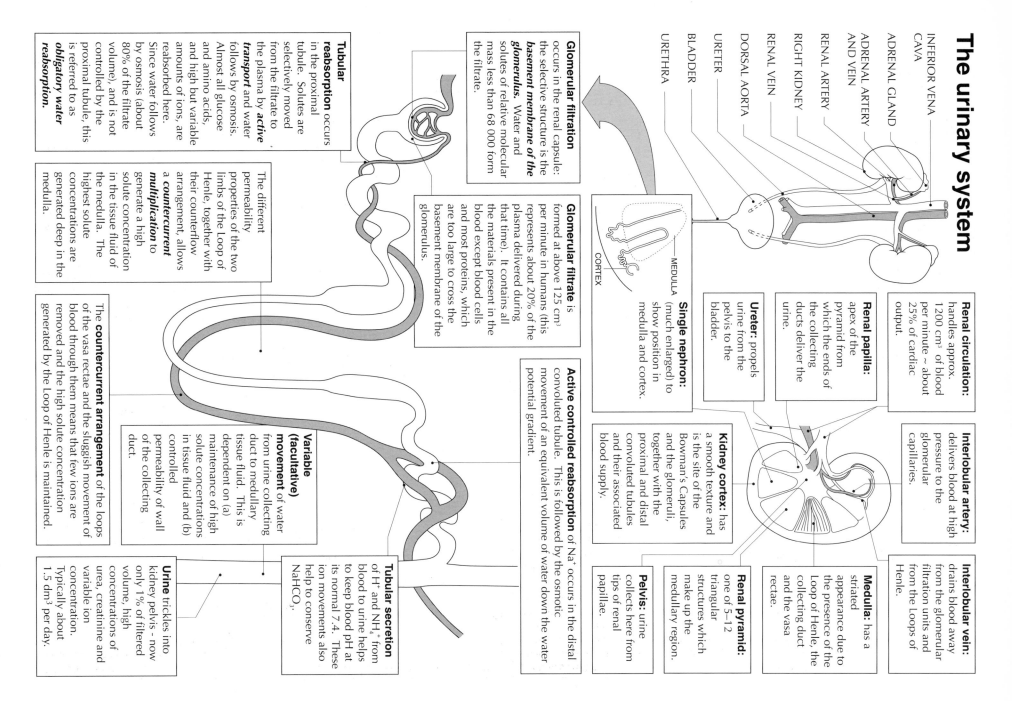

INFERIOR VENA CAVA

ADRENAL GLAND

ADRENAL ARTERY AND VEIN

RENAL ARTERY

RIGHT KIDNEY

RENAL VEIN

DORSAL AORTA

URETER

BLADDER

URETHRA

CORTEX

MEDULLA

Glomerular filtration occurs in the renal capsule: the selective structure is the *basement membrane of the glomerulus*. Water and solutes of relative molecular mass less than 68 000 form the filtrate.

Glomerular filtrate is formed at above 125 cm³ per minute in humans (this represents about 20% of the plasma delivered during that time). It contains all the materials present in the blood except blood cells and most proteins, which are too large to cross the basement membrane of the glomerulus.

Single nephron: (much enlarged) to show position in medulla and cortex.

Renal papilla: apex of the pyramid from which the collecting ducts deliver the urine.

Ureter: propels urine from the pelvis to the bladder.

Renal circulation: handles approx. 1 200 cm³ of blood per minute ~ about 25% of cardiac output.

Interlobular artery: delivers blood at high pressure to the glomerular capillaries.

Kidney cortex: has a smooth texture and is the site of the Bowman's Capsules and the glomeruli, together with the proximal and distal convoluted tubules and their associated blood supply.

Interlobular vein: drains blood away from the glomerular filtration units and from the Loops of Henle.

Medulla: has a striated appearance due to the presence of the Loop of Henle, the collecting duct and the vasa rectae.

Renal pyramid: one of 5–12 triangular structures which make up the medullary region.

Pelvis: urine collects here from tips of renal papillae.

Tubular reabsorption occurs in the proximal tubule. Solutes are selectively moved from the filtrate to the plasma by *active transport* and water follows by osmosis. Almost all glucose and amino acids, and high but variable amounts of ions, are reabsorbed here. Since water follows by osmosis (about 80% of the filtrate volume), and is not controlled by the proximal tubule, this is referred to as *obligatory water reabsorption*.

The different permeability properties of the two limbs of the Loop of Henle, together with their counterflow arrangement, allows *a countercurrent multiplication* to generate a high solute concentration in the tissue fluid of the medulla. The highest solute concentrations are generated deep in the medulla.

The **countercurrent arrangement** of the loops of the vasa rectae and the sluggish movement of blood through them means that few ions are removed and the high solute concentration generated by the Loop of Henle is maintained.

Variable (facultative) movement of water from urine collecting duct to medullary tissue fluid. This is dependent on (a) maintenance of high solute concentrations in tissue fluid and (b) controlled permeability of wall of the collecting duct.

Active controlled reabsorption of Na⁺ occurs in the distal convoluted tubule. This is followed by the osmotic movement of an equivalent volume of water down the water potential gradient.

Tubular secretion of H⁺ and NH₄⁺ from blood to urine helps to keep blood pH at its normal 7.4. These ion movements also help to conserve NaHCO₃.

Urine trickles into kidney pelvis - now only 1% of filtered volume, high concentrations of urea, creatinine and variable ion concentration. Typically about 1.5 dm³ per day.

Homeostasis: endocrine control of kidney action

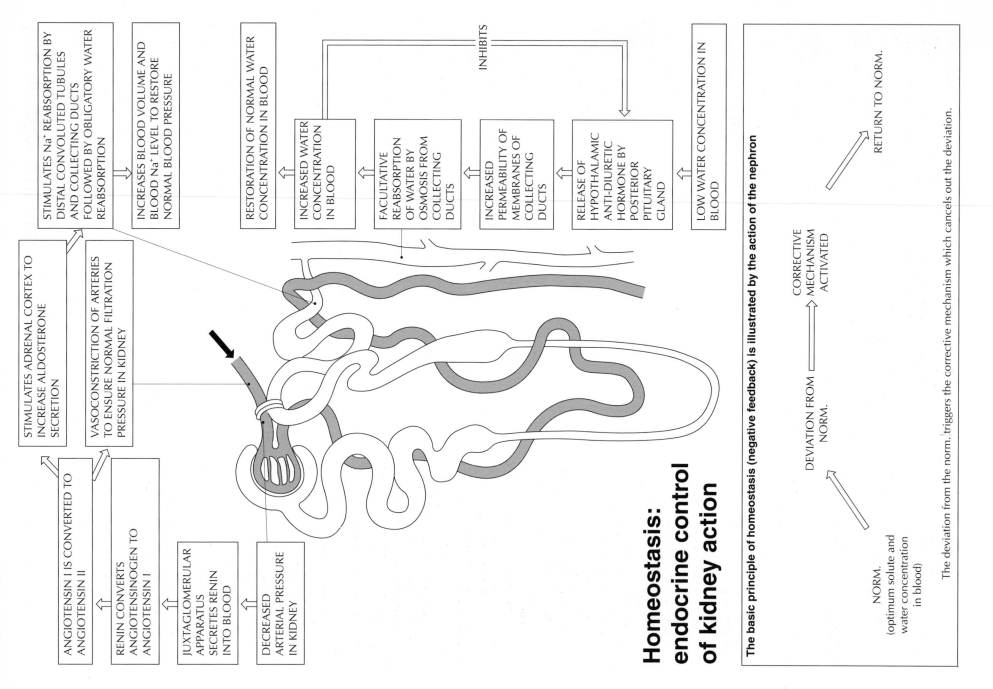

STIMULATES ADRENAL CORTEX TO INCREASE ALDOSTERONE SECRETION

STIMULATES Na⁺ REABSORPTION BY DISTAL CONVOLUTED TUBULES AND COLLECTING DUCTS FOLLOWED BY OBLIGATORY WATER REABSORPTION

INCREASES BLOOD VOLUME AND BLOOD Na⁺ LEVEL TO RESTORE NORMAL BLOOD PRESSURE

VASOCONSTRICTION OF ARTERIES TO ENSURE NORMAL FILTRATION PRESSURE IN KIDNEY

ANGIOTENSIN I IS CONVERTED TO ANGIOTENSIN II

RENIN CONVERTS ANGIOTENSINOGEN TO ANGIOTENSIN I

JUXTAGLOMERULAR APPARATUS SECRETES RENIN INTO BLOOD

DECREASED ARTERIAL PRESSURE IN KIDNEY

INHIBITS

RESTORATION OF NORMAL WATER CONCENTRATION IN BLOOD

INCREASED WATER CONCENTRATION IN BLOOD

FACULTATIVE REABSORPTION OF WATER BY OSMOSIS FROM COLLECTING DUCTS

INCREASED PERMEABILITY OF MEMBRANES OF COLLECTING DUCTS

RELEASE OF HYPOTHALAMIC ANTI-DIURETIC HORMONE BY POSTERIOR PITUITARY GLAND

LOW WATER CONCENTRATION IN BLOOD

The basic principle of homeostasis (negative feedback) is illustrated by the action of the nephron

NORM.
(optimum solute and water concentration in blood)

DEVIATION FROM NORM.

CORRECTIVE MECHANISM ACTIVATED

RETURN TO NORM.

The deviation from the norm. triggers the corrective mechanism which cancels out the deviation.

Motor (efferent) neurone: the dendrites (antennae), axon (cable), synaptic buttons (contacts) are serviced and maintained by the cell body.

Nissl's granules (or 'chromatophilic substance' because they take up stain readily) represent a highly ordered **rough endoplasmic reticulum.** Proteins made here, and passed into the neuronal processes (especially the axon), include structural proteins of the neurone membrane, transport proteins such as the Na⁺/K⁺ pump and enzymes involved in neurotransmitter synthesis.

Neurofibrils are formed from microtubules and microfilaments. They offer support to the cell body and may be involved in the transport of materials throughout the neurone.

Axon collateral is a side branch of the axon which means that one cell may direct impulses to more than one effector.

Schwann cell (neurilemmocyte) is a glial cell which encircles the axon. When the two 'ends' of the Schwann cell meet, overlapping occurs which pushes the nucleus and cytoplasm to the outside layer.

SCHWANN CELL
CYTOPLASM
FORMS
NEURILEMMA

LAYERS OF SCHWANN
CELL MEMBRANE
FORM MYELIN SHEATH

SCHWANN CELL
AXON MEMBRANE
NEUROFIBRIL
AXOPLASM

OVERLAPPING
SCHWANN CELL

Neurilemma is found only around axons of the peripheral nervous system, i.e. typical sensory and motor neurones. The neurilemma plays a part in the regeneration of damaged nerves by forming a tubular sheath around the damaged area within which regeneration may occur (very rare in CNS).

Myelin sheath is composed of 20–30 layers of Schwann cell membrane. The high phospholipid content of the sheath offers electrical insulation → **saltatory impulse conduction.** Not complete until late childhood so infants often have slow responses/poor co-ordination.

Synaptic end bulb or **synaptic button** is important in nerve impulse conduction from one neurone to another or from a neurone to an effector. They contain membrane-enclosed sacs (**synaptic vesicles**) which store **neurotransmitters** prior to release and diffusion to the post-synaptic membrane.

Axon terminal

Axon is the communication route between the cell body and the axon terminals. There are two intracellular transport systems: **axoplasmic flow** is slow, unidirectional protoplasmic streaming which supplies new axoplasm for new or regenerating neurones; **axonal transport** is faster, bi-directional and via microtubules and microfilaments. Axonal transport returns materials to the cell body for degradation/recycling *but* is the route taken by the **herpes virus** and the **rabies virus** to the cell body, where they multiply and cause their damage. The toxin produced by the **tetanus bacterium** uses the same route to reach the central nervous system.

Nodes of Ranvier are unmyelinated segments of the neurone. Since these are uninsulated, ion movements may take place which effectively lead to action potentials 'leaping' from one node to another during **saltatory conduction.**

Axon hillock is the point on the neuronal membrane at which a **threshold stimulus** may lead to the initiation of an **action potential.**

Dendrites are extensions of the cell body containing all typical cell body organelles. They provide a large surface area to receive information which they then pass on towards the cell body. The plasma has a high density of **chemically gated ion channels,** important in impulse transmission.

Cell body contains a well-developed nucleus and nucleolus and many organelles such as lysosomes and mitochondria. Many neurones also contain yellowish-brown granules of **lipofuscin pigment,** which may be a by-product of lysosomal activity and which increases in concentration as the neurone ages. There is **no mitotic apparatus** (centriole/spindle) in neurones more than six months old, which means that damaged neurones can never be replaced (although they may regenerate – see below).

Action potential: a depolarization of about 110 mV resulting from *an inward flow of Na⁺ ions*

An **action potential** is the depolarization–repolarization cycle at the neurone membrane following the application of a threshold stimulus. Since the depolarization–repolarization depend upon ion concentration gradients and upon time of ion channel opening, both of which are effectively fixed, *all action potentials are of the same size.* Thus a nerve cell obeys the *all-or-nothing principle:* if a stimulus is strong enough to generate an action potential, the impulse is conducted along the entire neurone *at a constant and maximum strength* for the existing conditions.

Depolarization: the voltage-gated sodium channels open so that Na⁺ ions can move *into the axon.*

a.   down a *Na⁺ concentration gradient*

b.   down an *outside–inside electrical gradient*

Na⁺

K⁺ K⁺

The inward movement of Na⁺ ions during depolarization is an example of a *positive feedback system.* As Na⁺ ions continue to move inward depolarization increases, which opens more sodium channels so more Na⁺ ions enter causing more depolarization and so on.

Return to resting potential: although the action potential involves Na⁺ and K⁺ movements, the changes in absolute ion concentrations are very small (probably no more than 1 in 10⁷). Many action potentials could be transmitted before concentration gradients are significantly changed – the sodium–potassium pump can quickly restore resting ion concentration gradients to pre–impulsive levels.

MEMBRANE POTENTIAL / mV

+40

0

−70

0   1   2   3

TIME / ms

Threshold value: any stimulus strong enough to initiate an impulse is called a *threshold* or *liminal stimulus.* The stimulus begins the depolarization of the neuronal membrane – once a sufficient number of voltage-gated sodium channels is opened, positive feedback will ensure a complete depolarization. Any stimulus weaker than a threshold stimulus is called a *sub-threshold* or *subliminal stimulus.* Such a stimulus is incapable of initiating an action potential, but a series of such stimuli *may* exert a cumulative effect which may be sufficient to initiate an impulse. This is the phenomenon of *summation of impulses.*

Repolarization: sodium channels are closed but potassium channels are open so that K⁺ ions are able to move *out of the axon.*

a.   down a K⁺ *concentration gradient*

b.   down an *electrochemical gradient*

Na⁺   Na⁺

K⁺

Hyperpolarization and refractory period: Potassium channels close and short term conformational changes in the pore proteins of the sodium channels mean that these voltage-gated sodium channels are *inactivated.* As a result the neuronal membrane becomes *refractory* – unable to respond to a stimulus which would normally trigger an action potential.

Na⁺   Na⁺

K⁺ K⁺

Synapse: structure and function

Transmission of an action potential across a chemical synapse involves a uni-directional release of molecules of neurotransmitter from pre-synaptic to post-synaptic membranes

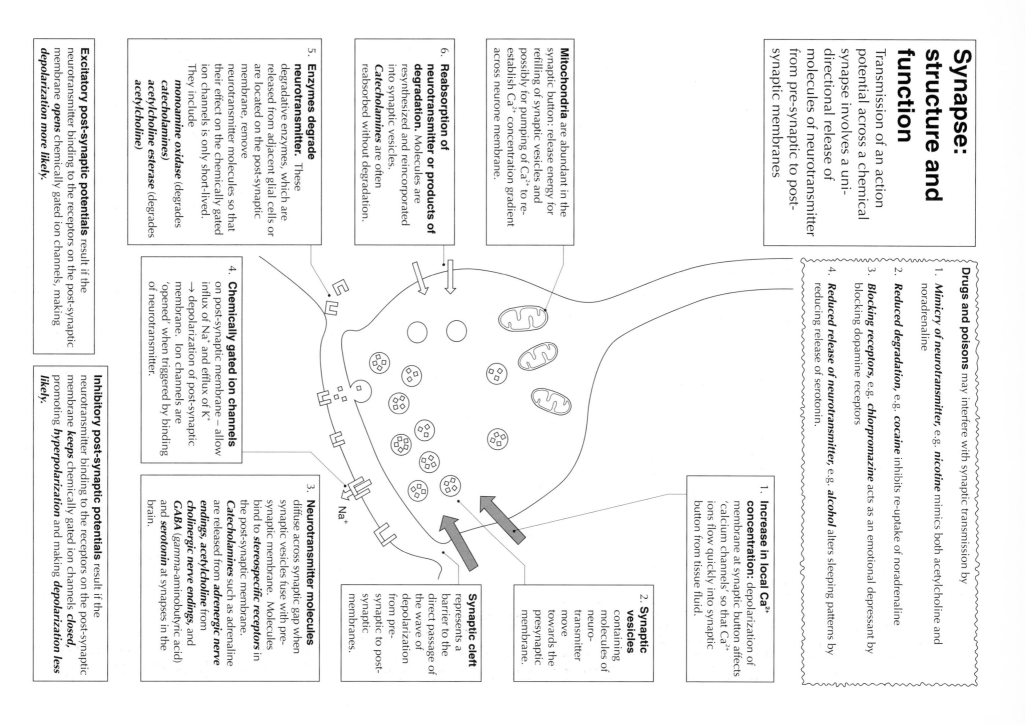

Drugs and poisons may interfere with synaptic transmission by

1. *Mimicry of neurotransmitter*, e.g. *nicotine* mimics both acetylcholine and noradrenaline

2. *Reduced degradation*, e.g. *cocaine* inhibits re-uptake of noradrenaline

3. *Blocking receptors*, e.g. *chlorpromazine* acts as an emotional depressant by blocking dopamine receptors

4. *Reduced release of neurotransmitter*, e.g. *alcohol* alters sleeping patterns by reducing release of serotonin.

Mitochondria are abundant in the synaptic button: release energy for refilling of synaptic vesicles and possibly for pumping of Ca^{2+} to re-establish Ca^{2+} concentration gradient across neurone membrane.

---

6. **Reabsorption of neurotransmitter or products of degradation.** Molecules are resynthesized and reincorporated into synaptic vesicles. *Catecholamines* are often reabsorbed without degradation.

---

5. **Enzymes degrade neurotransmitter.** These degradative enzymes, which are released from adjacent glial cells or are located on the post-synaptic membrane, remove neurotransmitter molecules so that their effect on the chemically gated ion channels is only short-lived. They include
monoamine oxidase (degrades *catecholamines*)
acetylcholine esterase (degrades *acetylcholine*)

---

4. **Chemically gated ion channels** on post-synaptic membrane – allow influx of Na$^+$ and efflux of K$^+$ → depolarization of post-synaptic membrane. Ion channels are 'opened' when triggered by binding of neurotransmitter.

---

3. **Neurotransmitter molecules** diffuse across synaptic gap when synaptic vesicles fuse with pre-synaptic membrane. Molecules bind to *stereospecific receptors* in the post-synaptic membrane. *Catecholamines* such as adrenaline are released from *adrenergic nerve endings*, *acetylcholine* from *cholinergic nerve endings*, and *GABA* (gamma-aminobutyric acid) and *serotonin* at synapses in the brain.

Synaptic cleft represents a barrier to the direct passage of the wave of depolarization from pre-synaptic to post-synaptic membranes.

---

1. **Increase in local Ca^{2+} concentration:** depolarization of membrane at synaptic button affects 'calcium channels' so that Ca^{2+} ions flow quickly into synaptic button from tissue fluid.

---

2. **Synaptic vesicles** containing molecules of neuro-transmitter move towards the presynaptic membrane.

Excitatory post-synaptic potentials result if the neurotransmitter binding to the receptors on the post-synaptic membrane **opens** chemically gated ion channels, making *depolarization more likely*.

Inhibitory post-synaptic potentials result if the neurotransmitter binding to the receptors on the post-synaptic membrane **keeps** chemically gated ion channels **closed**, promoting **hyperpolarization** and making *depolarization less likely*.

Propagation of action potentials

This sector is in its **resting state** (at **resting potential**) with the voltage-gated sodium channels **closed but active.**

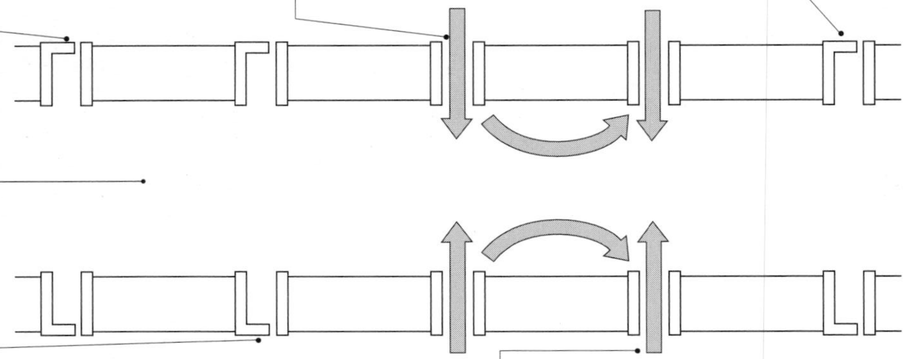

In this sector the voltage-gated sodium channels are **open** so that Na+ ions can move into the axon and the neuronal membrane becomes **depolarised.** Positive feedback and K+ ion outflow cause an action potential which begins to depolarise sodium channels if they are in the active state.

In **continuous conduction** an impulse is transmitted as a **wave of depolarisation** in which action potentials can be detected in immediately adjacent sectors of **unmyelinated neurones**

Local depolarisations begin to open voltage-gated sodium channels which are in the **closed but active** state. This can initiate an action potential in this sector.

In this sector the voltage-gated sodium channels are **closed but active:** the neuronal membrane is no longer in its **refractory period.**

In this sector the voltage-gated sodium channels are **closed and inactive.** The neuronal membrane is within its **refractory period** – it is unable to respond to depolarisation in adjacent sectors. This ensures that the waves of depolarisation – the **impulse** – can only travel in **one direction.**

NERVE CELL MEMBRANE

AXOPLASM

Direction of propagation of action potential

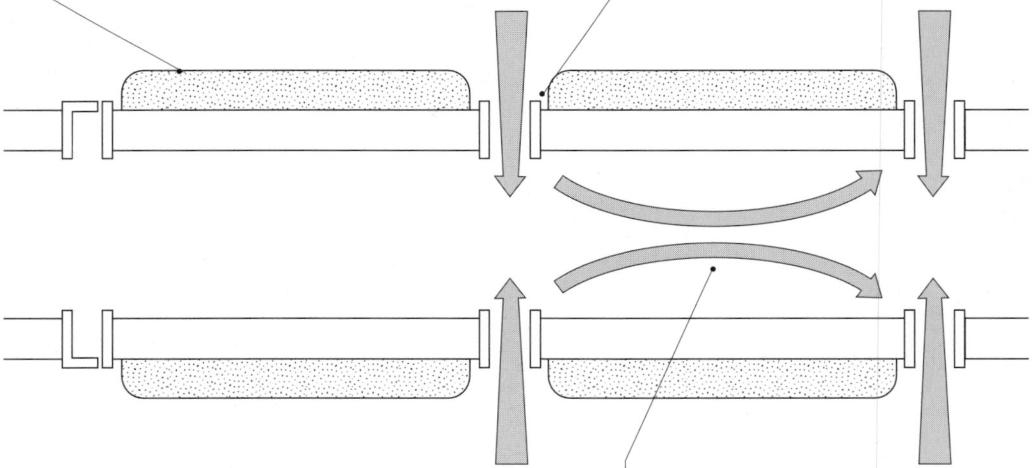

In **saltatory conduction** the impulse jumps from node to node of **myelinated neurones.**

Node of Ranvier is a region of unmyelinated nerve cell membrane. Voltage gated ion channels may allow Na+ and K+ movement across membrane so that an **action potential** can be generated.

Depolarisations of voltage-gated sodium channels are effective over greater distances i.e. between adjacent Nodes of Ranvier.

Myelin sheath contains **myelin**, a phospholipid which does not conduct an electric current. There are no transmembrane ion movements across myelinated membranes.

Alteration of synaptic transmission

The involvement of neurotransmitter chemicals, and the need to synthesise and degrade them, means that synaptic transmission is very prone to interference by both natural and artificial agents.

ALKALOSIS (increase in pH above 7.35 to about 8.0) causes increased excitability of neurones which may cause cerebral convulsions whilst **acidosis** (decrease in pH below 7.35 down to about 6.80) depresses neuronal activity and may lead to a comatose state.

REDUCTION IN RELEASE OF NEUROTRANSMITTER

Botulinus toxin causes paralysis by preventing release of **acetylcholine** – the toxin is very potent even in very small amounts (less than 0.0001 mg. may be fatal to a human) and is the substance responsible for one type of food poisoning. **Narcotic analgesics** such as morphine and heroin block the release of **substance P** from pain-transmitting neurones. **LSD**, an **hallucinogenic drug**, may inhibit release of **serotonin** (together with an increased secretion of adrenaline this disrupts the ability of the brain stem to "filter" information entering the cerebral cortex). **Alcohol** may cause a change in sleeping patterns, particularly increasing the frequency of waking periods, by reducing the synthesis and release of **serotonin**.

LIMITED REMOVAL OF NEUROTRANSMITTER

Catecholamines such **adrenaline** and **nor-adrenaline** are removed from the synaptic gap, and therefore from post-synaptic receptors, by re-uptake into the synaptic button. **Cocaine** acts as a **euphoric drug** by inhibiting the re-uptake of nor-adrenaline. **Strychnine and organophosphorous weedkillers/insecticides** inhibit **acetylcholinesterase** and so prolong the effects of one discharge of **acetylcholine**. **Amphetamines** inhibit the enzyme **monoamine oxidase**, which degrades the catecholamines **adrenaline** and **dopamine**, and thus act as **stimulants**.

MIMICRY OF NEUROTRANSMITTER

Nicotine, at low concentrations, mimics **acetylcholine** at nerve endings in the parasympathetic nervous system (N.B. Effect non-specific so heart racing and raised blood pressure result from stimulation of sympathetic branch!). **Muscarine**, a fungal hypnotic, also induces euphoria by mimicking acetylcholine in the parasympathetic nervous system. **Endorphins** such as β–**endorphin** and **enkephalin** are endogenously produced peptides which mimic the **opiate drugs** such as **heroin** and **morphine** and act as **natural painkillers** and **euphorics**.

DECREASED SENSITIVITY OF POST-SYNAPTIC MEMBRANE is the function of many **depressants**.

Barbiturates reduce the ion-permeability of the post-synaptic membrane by preferentially binding to Na^+ or K^+ ions (e.g. Barbital + Na^+ → sodium barbital + H^+).

INHIBITION OF ADENYL CYCLASE. This enzyme degrades the "second messenger" cyclic AMP, responsible for post-synaptic activity in some neurones. **Caffeine** and **theophylline** both inhibit the enzyme and thus artificially maintain cAMP levels acting as mild **stimulants**.

ALTERED AVAILABILITY OF POST-SYNAPTIC RECEPTORS

Curare, a poison used by South American Indians to tip the heads of arrows and darts, competes for **acetylcholine receptors**. Since the muscles of respiration depend on neuromuscular transmission to initiate their contraction, death results from asphyxiation. Some **hypnotic drugs** increase the number of post-synaptic receptors – each receptor is less likely to be affected by the same "charge" of neurotransmitter and the post-synaptic system is generally depressed. Very high doses of **nicotine** prevent binding of **acetylcholine** to appropriate receptors, leading to death. **Major tranquillizers** such as **chlorpromazine** block **dopamine receptors** in the brain and thus depress emotional responses.

AUTOIMMUNE RESPONSE in which the body produces antibodies to **acetylcholine receptors** is the cause of the disease **myasthenia gravis**, characterised by muscular weakness. **Anticholinesterase drugs** such as **neostigmine** (see above) may improve acetylcholine-receptor binding, and **steroid drugs** such as **prednisone** may reduce antibody levels.

The eye as a sense organ

has a *generator region* (the retina) and a range of *ancillary structures* (e.g. lens, iris and choroid) to ensure optimum operation of the sensory cells.

Sclera: a tough coat of collagen fibres which protects the eyeball against mechanical damage, helps to maintain the shape of the eyeball, and provides attachment for the tendons of the rectus muscles.

Choroid: a thin pigmented layer which absorbs light to prevent internal reflection and multiple image formation. It has a well developed blood vascular supply which services the cells of the retina.

Rectus muscle: one of three pairs of striated muscles which adjust the position of the eyeball within the orbit (eye socket).

Ciliary body: contains both circular and radial muscles which are able to alter the shape of the lens. The ciliary body also supports the lens behind the pupil, and secretes the aqueous humour into the anterior chamber of the eye.

Aqueous humour: maintains the curvature of the choroid.

Conjunctiva: protects the cornea at the front of the eyeball against friction and minor mechanical damage. Cells are replaced very rapidly.

Cornea: transparent to admit light to anterior chamber. Most light refraction occurs at the cornea/aqueous humour boundary.

Pupil: the circular opening which admits light to the lens.

Canal of Schlemm: a sinus which contains venous drainage products of the sclera and choroid.

Iris: a pigmented muscular structure which controls the entry of light via the pupil. Has both radial and circular muscles – there is a well developed cranial reflex which alters the diameter of the pupil to ensure optimum illumination of the retina. Degree of pigmentation (by melanin) is genetically determined.

Retina: contains light-sensitive cells, the rods and cones, and a series of neurones which enhance image formation and transmit action potentials to the optic nerve.

Fovea centralis/yellow spot: region in which only cones (no rods) are found and thus the area of greatest visual acuity.

Blind spot: this region contains no light-sensitive cells and thus an image falling on this area cannot be perceived. The blind spot corresponds to the exit of the axons of the ganglion cells of the optic nerve.

Optic nerve: transmits impulses generated in the retina to the visual cortex of the cerebral cortex.

Retinal blood vessels: delivery of nutrients/removal of waste products to/from cells of retina.

Vitreous humour: this is produced during embryonic life and never replaced. It helps to maintain the shape of the eyeball, supports the lens and keeps the retina firmly applied to the choroid. There is considerable refraction of light at the posterior edge of the lens where it meets the vitreous humour.

Lens: is normally completely transparent. It allows light to enter the posterior chamber of the eye and is responsible for the refraction necessary to complete the fine focusing of an image on to the retina. The lens is composed of numerous layers of protein fibres in an elastic capsule.

Suspensory ligaments: attach the lens capsule to the muscular ciliary process, thus permitting adjustment of the lens shape during accommodation.

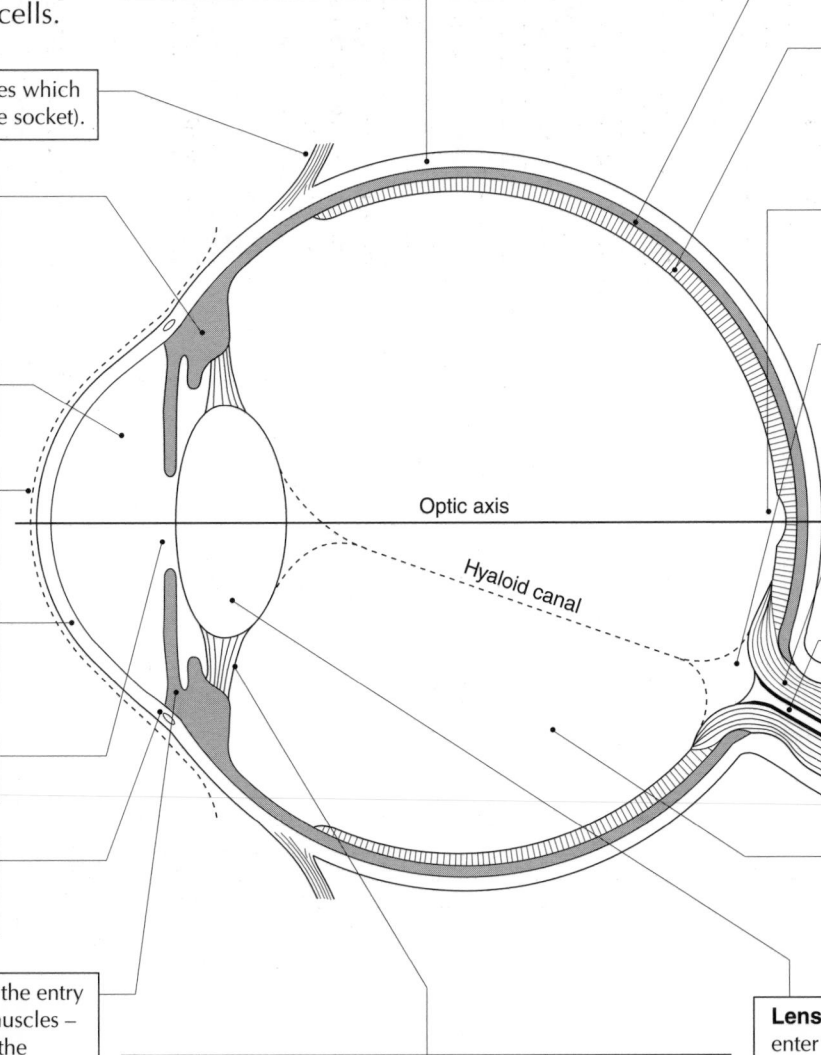

Optic axis

Hyaloid canal

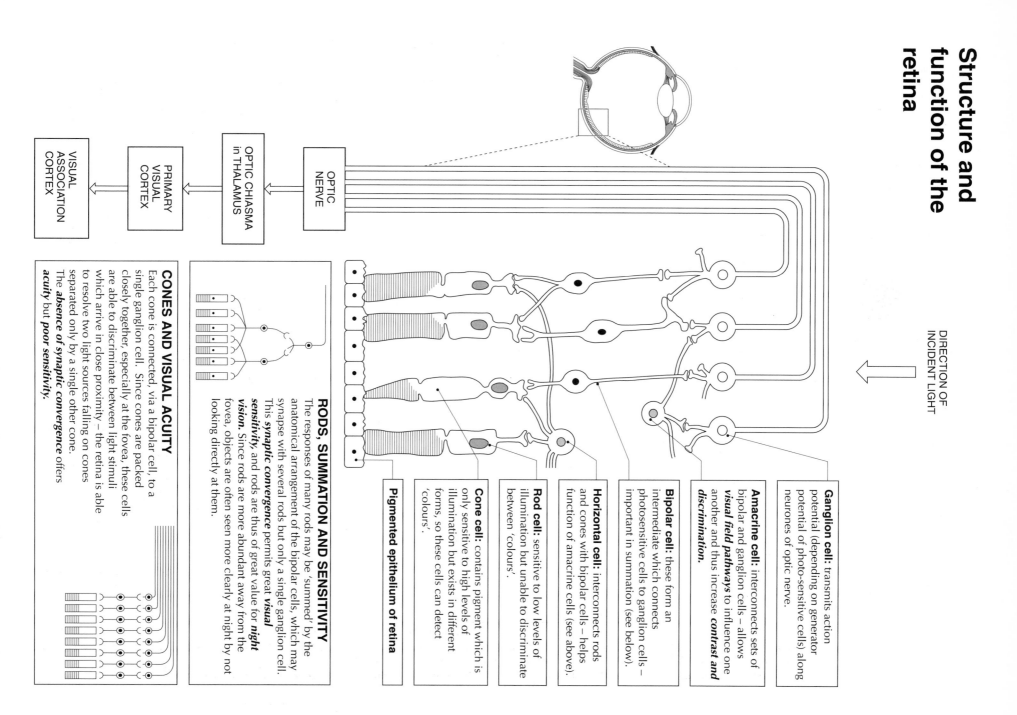

DIRECTION OF
INCIDENT LIGHT

OPTIC
NERVE

OPTIC CHIASMA
in THALAMUS

PRIMARY
VISUAL
CORTEX

VISUAL
ASSOCIATION
CORTEX

Ganglion cell: transmits action potential (depending on generator potential of photo-sensitive cells) along neurones of optic nerve.

Amacrine cell: interconnects sets of bipolar and ganglion cells – allows *visual field pathways* to influence one another and thus increase *contrast and discrimination.*

Bipolar cell: these form an intermediate which connects photosensitive cells to ganglion cells – important in summation (see below).

Horizontal cell: interconnects rods and cones with bipolar cells – helps function of amacrine cells (see above).

Rod cell: sensitive to low levels of illumination but unable to discriminate between 'colours'.

Cone cell: contains pigment which is only sensitive to high levels of illumination but exists in different forms, so these cells can detect 'colours'.

Pigmented epithelium of retina

RODS, SUMMATION AND SENSITIVITY

The responses of many rods may be 'summed' by the anatomical arrangement of the rods but only a single ganglion cell, which may synapse with several rods but only a single ganglion cell. This *synaptic convergence* permits great *visual sensitivity*, and rods are thus of great value for *night vision*. Since rods are more abundant away from the fovea, objects are often seen more clearly at night by not looking directly at them.

CONES AND VISUAL ACUITY

Each cone is connected, via a bipolar cell, to a single ganglion cell. Since cones are packed closely together, especially at the fovea, these cells are able to discriminate between light stimuli which arrive in close proximity – the retina is able to resolve two light sources falling on cones separated only by a single other cone. The *absence of synaptic convergence* offers *acuity* but *poor sensitivity*.

Structure and function in rod cells

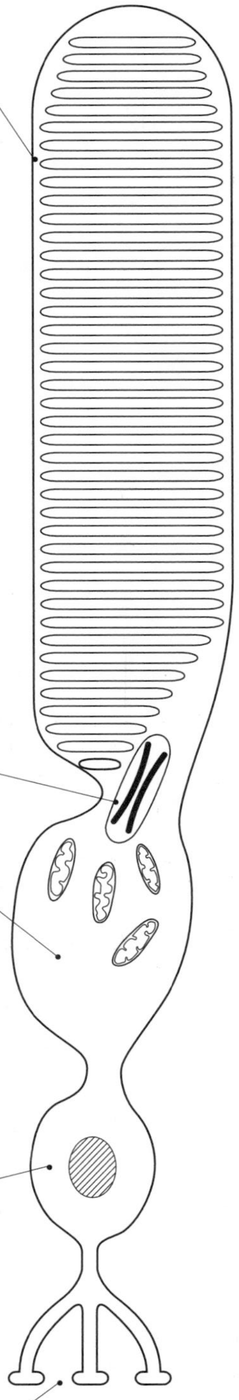

THE VISUAL CYCLE

The light-sensitive compound **rhodopsin** is a conjugate protein of **opsin** and **retinal** (a derivative of a vitamin A).

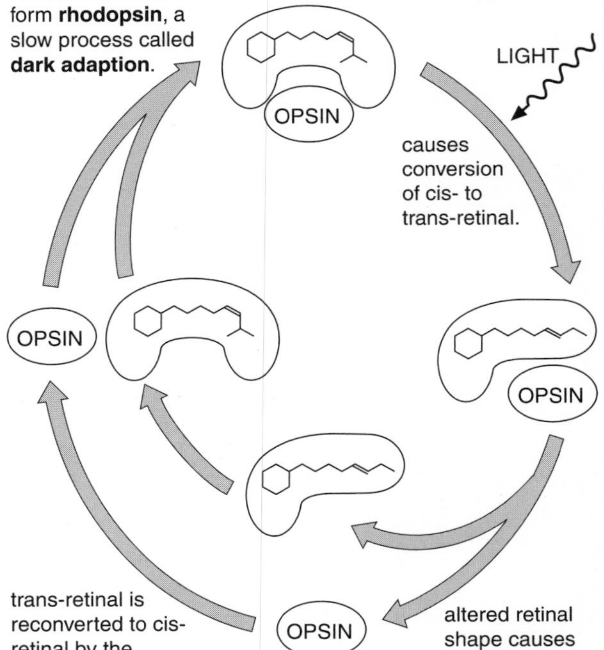

LIGHT

causes conversion of cis- to trans-retinal.

altered retinal shape causes separation from opsin, which undergoes a conformational change.

trans-retinal is reconverted to cis-retinal by the enzyme **retinal isomerase.**

Opsin and cis-retinal recombine to form **rhodopsin**, a slow process called **dark adaption.**

In the **light**, rhodopsin dissociates causing changes in intracellular Ca^{2+} concentration. This reduces the return of Na^+ to the outer segment and the inside of the rod cell becomes **hyperpolarised.**

In the **dark**, rhodopsin does not dissociate and there is an uninterrupted cycling of Na^+ between inner and outer segments via the extracellular fluid.

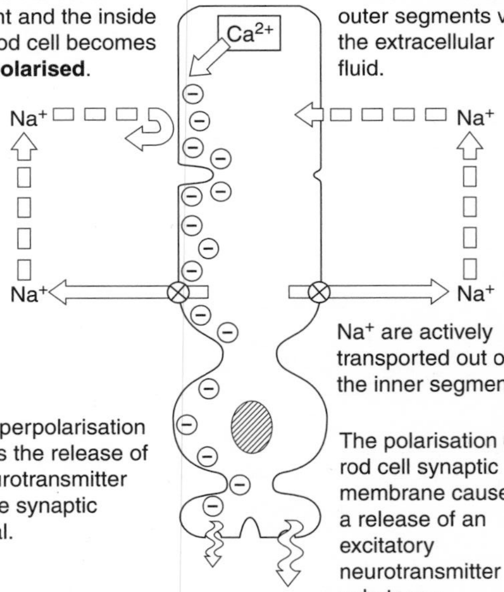

Na^+ are actively transported out of the inner segment.

The polarisation of rod cell synaptic membrane causes a release of an excitatory neurotransmitter substance.

This hyperpolarisation reduces the release of the neurotransmitter from the synaptic terminal.

The degree of excitation of the rod cell is signalled by the **degree of hyperpolarisation** of the rod cell membrane – this is the equivalent of an **amplitude modulated generator potential.** If the generator potential exceeds a threshold value there may be an action potential triggered in the ganglion cells of the optic nerve.

OUTER SEGMENT contains numerous discs or lamellae which are packed with the photosensitive compound **rhodopsin**. Light falling on rhodopsin induces a conformational change in the protein part of the molecule which triggers a cascade via the hydrolysis of cyclic GMP which eventually hyperpolarises the rod cell membrane.

NECK REGION contains a modified cilium and basal body, and connects the metabolic centre of the inner segment to the photo-sensitive region.

INNER SEGMENT contains numerous mitochondria (ATP for active transport of Na^+ ions), ribosomes (for synthesis of opsin) and an extensive endoplasmic reticulum (Ca^{2+} storage and membrane assembly).

CELL BODY with nucleus

SYNAPTIC TERMINAL releases an excitatory neurotrans-mitter. There is maximal release of this compound **in the dark.**

Cones and colour vision

OUTER SEGMENT

NECK PIECE

INNER SEGMENT

CELL BODY

SYNAPTIC TERMINAL

The visual cycle in cones is very similar to that in rods, except that the protein component of the photosensitive compound (**iodopsin** or **photopsin**) is different. In particular the dissociation of the pigment which leads to the generator potential requires a higher energy input – thus cones are not very sensitive in dim light.

Trichromatic theory of colour vision suggests that there are three variants of iodopsin, each of which is sensitive to a different range of wavelengths, corresponding to the three primary colours **red**, **blue** and **green**.

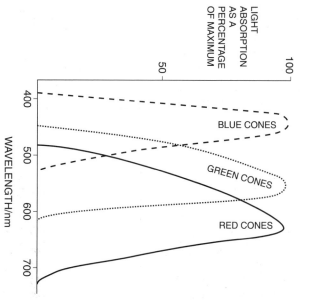

LIGHT ABSORPTION AS A PERCENTAGE OF MAXIMUM

100

50

WAVELENGTH/nm

400 · 500 · 600 · 700

BLUE CONES

GREEN CONES

RED CONES

Each type of iodopsin is probably located in different cones, and different colours are perceived in the brain from the sensory input from combinations of the three "primary cones".

| PRIMARY CONES STIMULATED | | | COLOUR PERCEIVED IN BRAIN |
BLUE (440 nm)	GREEN (550 nm)	RED (600 nm)	
+	+	+	WHITE
+			BLUE
	+		GREEN
		+	RED
+	+		CYAN
	+	+	ORANGE/YELLOW
+		+	MAGENTA

The trichromatic theory is supported by studies using recombinant DNA techniques which demonstrated separate genes for blue-sensitive, green-sensitive and red-sensitive opsin.

Endocrine control depends upon chemical messengers secreted from cells and binding to specific hormone receptors

A hormone is a chemical produced in one part of an organism which is transported throughout the organism and produces a specific response in target cells.

1 Stimulus affects endocrine gland so that it releases a hormone (chemical messenger).

2 Hormones are distributed throughout the body in the bloodstream.

3 Receptors on the membranes of the target organ 'recognise' the circulating hormone molecules. This mechanism ensures that only the specific target organ can respond to the hormone.

4 Target organ brings about an appropriate response to deal with the original stimulus.

NERVOUS AND CHEMICAL CO-ORDINATION COMPARED

	NERVOUS	CHEMICAL (ENDOCRINE)
NATURE OF MESSAGE	Electrochemical impulses	Chemical compounds (hormones)
ROUTE OF TRANSMISSION	Specific nerve cells	General blood system
TYPE OF EFFECTS	Rapid, but usually short-term e.g. blinking	Usually slower, but generally longer-lasting e.g. growth

Note that the hormone **adrenaline** has effects which are rapid but short-term, and therefore partially mimics some effects of the nervous system.

Pituitary gland secretes
- a number of hormones (trophins) which influence other endocrine organs
- human growth hormone
- anti diuretic hormone.

Thyroid gland secretes thyroxine which controls metabolic rate and is important in regulation of body temperature.

Adrenal glands secrete **adrenaline** which is called the **fight** or **flight hormone** because it prepares the body for action.

- more glycogen is converted to glucose
- deeper, more rapid breathing
- faster heartbeat
- diversion of blood from gut to muscles.

The above all help to provide more glucose and more oxygen for the working muscles. There may also be
- dilation of the pupils
- raising of the hair.

Testes secrete testosterone, which
- controls the production and development of sperm
- regulates the development of the male secondary sexual characteristics
i.e. enlargement of sex organs, growth of facial hair, muscle enlargement, deepening of voice, sexual behaviour

Artificial steroids increase muscle bulk but inhibit natural testosterone production which can lead to sterility.

The pancreas had both **endocrine** (secretes the hormones insulin and glucagon) and **exocrine** (secretes a range of digestive enzymes including lipase and amylase) functions.

HISTOLOGY OF PANCREAS

Islets of Langerhans (endocrine)

Acinar cells (exocrine)

Ovaries secrete the female sex hormones.
Oestrogen
- controls the development of the female secondary sexual characteristics, including the pattern of laying down of fat which gives the feminine body shape
- controls the repair of the lining of the uterus
- helps to regulate the development of further follicles.

Progesterone
- prepares the lining of the uterus for implantation by increasing the thickness of the lining and number of blood vessels
- helps to inhibit the growth of further follicles.

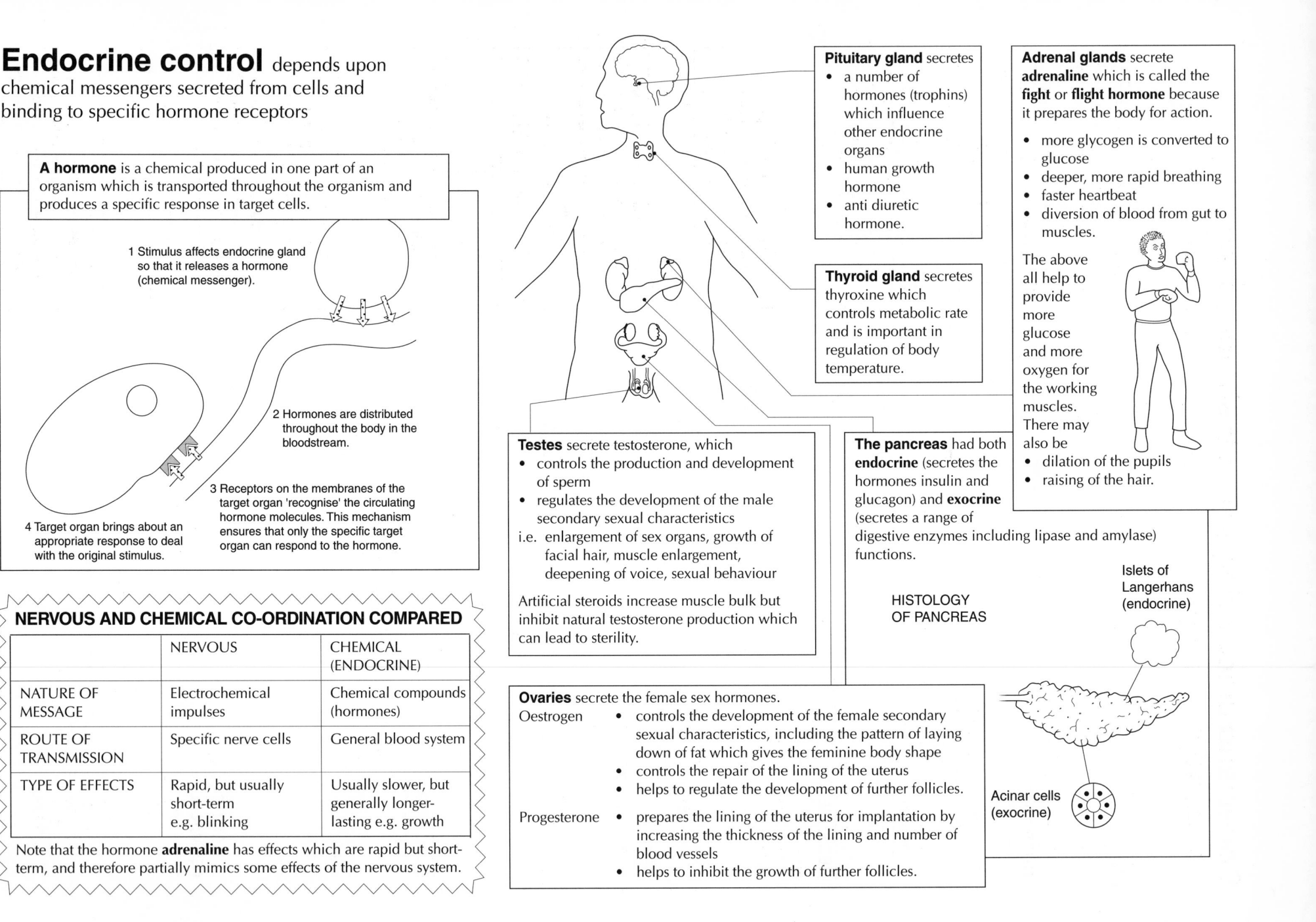

Some Hormones bring about their action via a second messenger which is commonly cyclic amp (cAMP)

NB.
1. The hormone **does not enter the cell.**

2. Each hormone-receptor complex may activate several molecules of G-proteins and hence several molecules of adenylate cyclase. Each adenylate cyclase produces many cAMP molecules, and each active protein kinase may phosphorylate many molecules of cellular protein. Thus the original hormone message is **amplified** by a **cascade effect.**

Hormones which exert their effects in this way include **anti-diuretic hormone, oxytocin, calcitonin, trophic hormones, adrenaline** and **glucagon.**

1. **Endocrine cell** produces and secretes **hormone, e.g. adrenaline.**

2. **Hormone** is transported through bloodstream. This hormone is the

   FIRST MESSENGER

3. Hormone binds to **specific protein receptor** in plasma membrane of target cell e.g. Adrenaline binds to **β-adrenergic receptors.**

4. **Hormone-receptor complex** binds to and activates (by promoting binding to GTP) a **G-protein** within the plasma membrane.

5. G-protein activates the enzyme **adenylate cyclase** which is located on the inner surface of the plasma membrane. This enzyme catalyses the reaction

   ATP → 3'5' cyclic AMP

   cAMP is the SECOND MESSENGER

6. cAMP promotes the activation (by **phosphorylation)** of an intracellular **protein kinase.**

7. Active protein kinase catalyses the phosphorylation of **specific cellular proteins** and in this way alters cellular activity e.g. adrenaline may initiate a series of reactions which end by **activating glycogen phosphorylase** and **inhibiting glycogen synthase** with the net result that GLYCOGEN → GLUCOSE.

8. cAMP is removed by a **phosphatase** enzyme. This enzyme may be inhibited by **caffeine** and **nicotine** which therefore prolong the effects of the original hormone

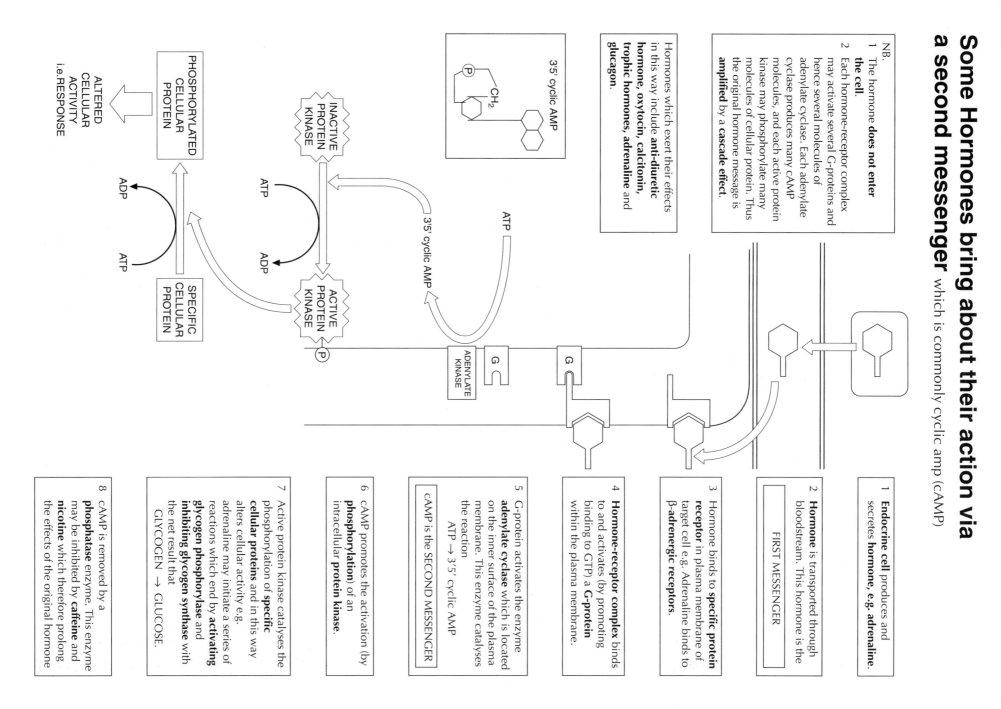

3'5' cyclic AMP

ADENYLATE KINASE

INACTIVE PROTEIN KINASE

ACTIVE PROTEIN KINASE

ATP

ADP

ATP

ADP

SPECIFIC CELLULAR PROTEIN

PHOSPHORYLATED CELLULAR PROTEIN

ALTERED CELLULAR ACTIVITY i.e. RESPONSE

ATP

CH₂

Spinal cord and reflex action

Reflex actions are rapid responses to internal or external stimuli which allow the body to maintain homeostasis. They may be **somatic** or **autonomic** but all involve the sequence

RECEPTOR → SENSORY NEURONE → CNS → MOTOR NEURONE → EFFECTOR

Dorsal root ganglion is a swelling in the dorsal root of the spinal nerve caused by an aggregation of cell bodies of the sensory neurones.

Receptor is the origin of the reflex action. The receptor responds to a **stimulus** (a change in the environment) by producing an amplitude-modulated generator potential which is transmitted along the sensory neurone as a **frequency-modulated action potential.**

Ascending fibre can transmit sensory input to the higher centres of the CNS (medulla, cerebellum and/or cerebral cortex).

Interneurone (associate or **internuncial neurone)** transmits sensory input across the spinal cord. It is not myelinated so that impulse transmission is relatively slow. This is significant in permitting higher centres to modify reflex action via descending fibres. This interneurone may be absent in the most rapid and inflexible reflexes, such as the knee jerk.

Central canal contains cerebrospinal fluid (CSF) which nourishes and maintains electrolyte balance in the CNS.

Grey matter is largely composed of cell bodies (mainly of motor neurones) and unmyelinated axons of interneurones. There are also many **glial cells** (support and nourish neurones). The absence of myelin is responsible for the grey colour of this region.

Sensory neurone transmits information from a receptor towards the CNS. Sensory neurones are myelinated so that transmission is rapid.

Spinal nerve is one of 31 pairs and is a **mixed** nerve, i.e. it has both sensory and motor fibres within in. Contains axons supported by connective tissue and fat.

Effectors may be muscles or glands and perform an action (response) when they receive an input from the motor neurone. This response will have a **survival value** to the organism in which it takes place.

Motor neurone of sympathetic nervous system transmits impulses to a variety of effectors other than striated (skeletal) muscle. Thus one sensory input can bring about a variety of responses.

Motor neurone of somatic nervous system transmits impulses from the CNS to an effector (typically striated muscle fibres). These neurones are myelinated and form the bulk of the ventral root of the spinal nerve.

White matter consists of bundles of motor and sensory neurones, including those of the ascending and descending tracts. These cells are well myelinated – hence the light colour of this region.

Descending fibre from the brain can transmit impulses from higher centres which may modify reflex action by influencing post-synaptic potentials in the outgoing motor neurones. Descending fibres may be **inhibitory** (minimize the reflex action) or **excitatory** (exaggerate the action), and both **temporal** and **spatial** summation of impulses from interneurones and descending fibres may take place. Both ascending and descending fibres are myelinated (medullated) so that they transmit impulses rapidly to and from the brain.

Synapse in the sympathetic ganglion of the autonomic nervous system has adrenaline or noradrenaline as the major neurotransmitter.

The human nervous system

can be subdivided into central and peripheral components

CENTRAL NERVOUS SYSTEM (CNS)

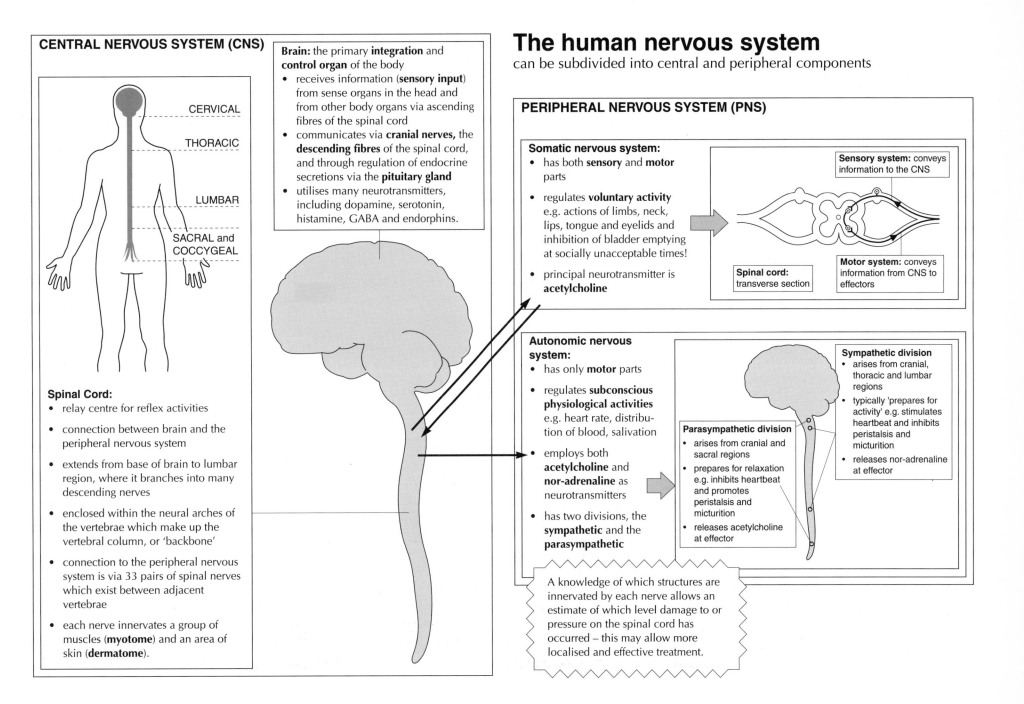

CERVICAL

THORACIC

LUMBAR

SACRAL and COCCYGEAL

Brain: the primary **integration** and **control organ** of the body

- receives information (**sensory input**) from sense organs in the head and from other body organs via ascending fibres of the spinal cord
- communicates via **cranial nerves,** the **descending fibres** of the spinal cord, and through regulation of endocrine secretions via the **pituitary gland**
- utilises many neurotransmitters, including dopamine, serotonin, histamine, GABA and endorphins.

Spinal Cord:

- relay centre for reflex activities

- connection between brain and the peripheral nervous system

- extends from base of brain to lumbar region, where it branches into many descending nerves

- enclosed within the neural arches of the vertebrae which make up the vertebral column, or 'backbone'

- connection to the peripheral nervous system is via 33 pairs of spinal nerves which exist between adjacent vertebrae

- each nerve innervates a group of muscles (**myotome**) and an area of skin (**dermatome**).

PERIPHERAL NERVOUS SYSTEM (PNS)

Somatic nervous system:

- has both **sensory** and **motor** parts

- regulates **voluntary activity** e.g. actions of limbs, neck, lips, tongue and eyelids and inhibition of bladder emptying at socially unacceptable times!

- principal neurotransmitter is **acetylcholine**

Sensory system: conveys information to the CNS

Spinal cord: transverse section

Motor system: conveys information from CNS to effectors

Autonomic nervous system:

- has only **motor** parts

- regulates **subconscious physiological activities** e.g. heart rate, distribution of blood, salivation

- employs both **acetylcholine** and **nor-adrenaline** as neurotransmitters

- has two divisions, the **sympathetic** and the **parasympathetic**

Parasympathetic division

- arises from cranial and sacral regions
- prepares for relaxation e.g. inhibits heartbeat and promotes peristalsis and micturition
- releases acetylcholine at effector

Sympathetic division

- arises from cranial, thoracic and lumbar regions
- typically 'prepares for activity' e.g. stimulates heartbeat and inhibits peristalsis and micturition
- releases nor-adrenaline at effector

A knowledge of which structures are innervated by each nerve allows an estimate of which level damage to or pressure on the spinal cord has occurred – this may allow more localised and effective treatment.

Structure and contraction of striated (skeletal) muscle

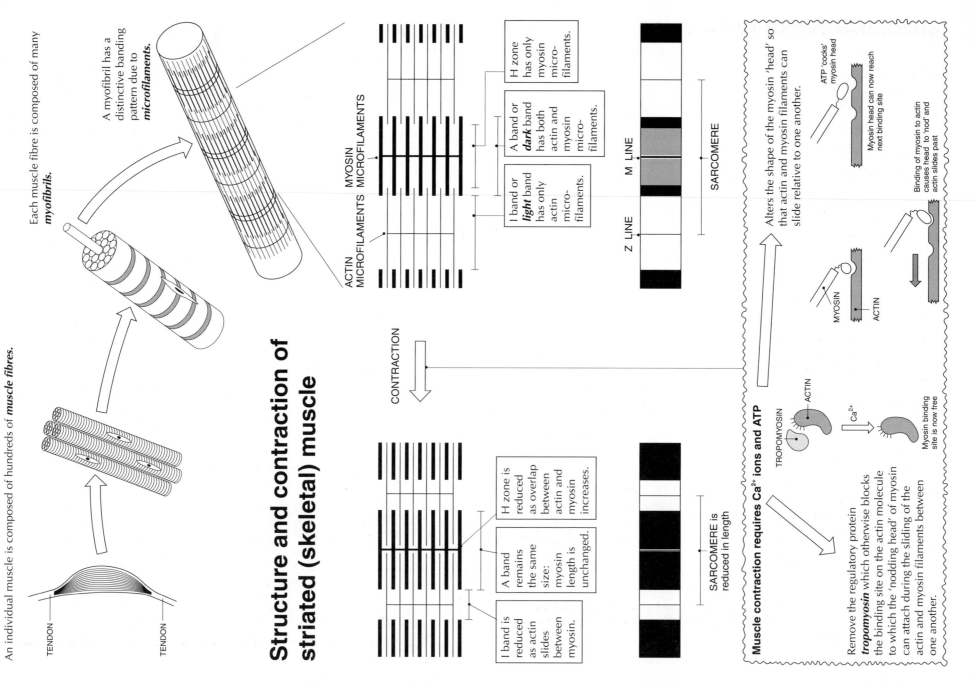

An individual muscle is composed of hundreds of *muscle fibres.*

TENDON

TENDON

Each muscle fibre is composed of many *myofibrils.*

A myofibril has a distinctive banding pattern due to *microfilaments.*

MYOSIN MICROFILAMENTS

ACTIN MICROFILAMENTS

H zone has only myosin microfilaments.

A band or *dark* band has both actin and myosin microfilaments.

I band or *light* band has only actin microfilaments.

Z LINE

M LINE

SARCOMERE

CONTRACTION

H zone is reduced as overlap between actin and myosin increases.

A band remains the same size: myosin length is unchanged.

I band is reduced as actin slides between myosin.

SARCOMERE is reduced in length

Alters the shape of the myosin 'head' so that actin and myosin filaments can slide relative to one another.

ATP 'cocks' myosin head

Myosin head can now reach next binding site

Binding of myosin to actin causes head to 'nod' and actin slides past

MYOSIN

ACTIN

Muscle contraction requires Ca²⁺ ions and ATP

TROPOMYOSIN

ACTIN

Ca²⁺

Myosin binding site is now free

Remove the regulatory protein *tropomyosin* which otherwise blocks the binding site on the actin molecule to which the 'nodding head' of myosin can attach during the sliding of the actin and myosin filaments between one another.

122 Striated muscle

Muscles: effects of exercise

SHORT TERM

Blood flow to muscles may increase by up to 25 times during exercise.

Respiration and oxygen debt: once the creatine phosphate supply has been exhausted ATP may be generated by anaerobic respiration: exercise may generate an oxygen debt of 10–12 dm^3 (up to 18–20 dm^3 in trained athletes).

Fatigue and exhaustion: "Fatigue" is the inability to repeat muscular contraction with the same force and is at least partially explained by toxic effects of accumulated lactate.

Eventually fatigue gives way to exhaustion, which is associated with depleted K^+ concentration in muscle cells.

Damage: for example tearing or straining through over-stretching without warming-up.

Heavy exercise leads to shorter, tighter muscles which are more prone to such damage.

Muscle soreness following exercise is due to minor inflammation and associated tissue swelling during the recovery period

Cramp is a powerful, sustained and uncontrolled contraction resulting from over-exercise, especially of muscles with inadequate circulation.

It is made worse by lackate, chilling and low concentrations of oxygen or salt.

Glycogen and potassium depletion: these are associated with fatigue and exhaustion. Several days of recovery may be necessary to return to optimum concentrations of these solutes.

Glycogen can be an energy 'store'. This compound can be 'loaded' into muscles by following a period of carbohydrate starvation (48h.) with a high intake of simple sugars (24h.) – useful marathon preparation!

LONG TERM

Muscle size is genetically determined; high testosterone concentration (anabolic steriod) increases muscle growth.

Exercise may increase muscle size by up to 60%, mainly by increased diameter of individual fibres.

Co-ordination: improved response, especially between antagonistic pairs, since as prime mover increases its speed of contraction the antagonist must be allowed to relax more quickly.

Biochemical changes associated with an increase in the number and size of the mitochondria is an increased activity of enzymes of the TCA cycle, electron transport and fatty acid oxidation (leading to a doubling of mitochondrial efficiency).

Stores of creatine phosphate, glycogen, fat and myoglobin are doubled and there is a more rapid release of fatty acids from the stores.

Blood supply: number of vessels and extent of capillary beds increases, both to improve *delivery of substrates* and *clearance of toxic products*

SPRINT OR MARATHON? FAST AND SLOW MUSCLE FIBRES

	FAST (WHITE)	SLOW (RED)
STRUCTURE	Few mitochondria	Many mitochondria
	Little myoglobin	Much myoglobin
	Much glycogen	Little glycogen
LOCATION	Relatively superficial	Deep-seated within limbs
GENERAL PROPERTIES	Relatively excitable – 'all-or-nothing' response	Low excitability – 'graded' response
	Fast contraction	Slow contraction
	Fatigues quickly	Fatigues slowly
	Rapid formation of oxygen debt–anaerobic	Respiration mainly aerobic – **can** build up oxygen debt
FUNCTION	Immediate, fast contraction: sprint	Sustained contraction: marathon, maintenance of posture

Section of seminiferous tubule

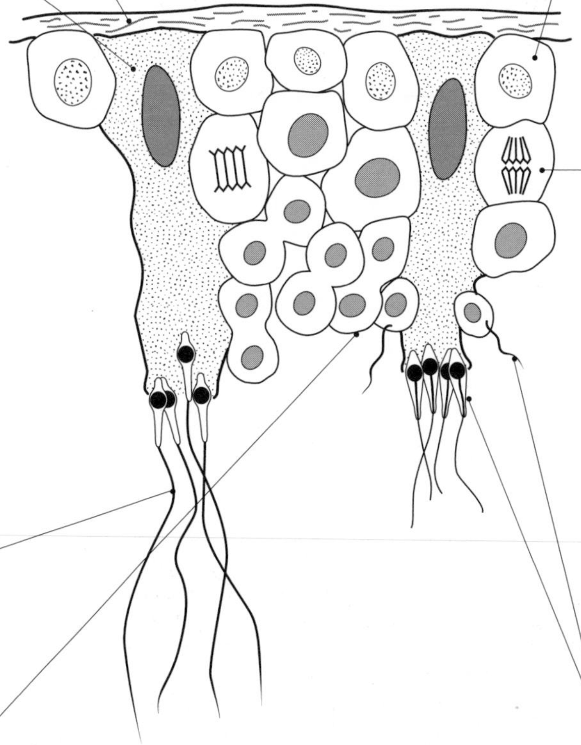

BASEMENT MEMBRANE

SERTOLI CELL is joined to other Sertoli cells by junctions which form a blood-testis barrier which allows the spermatozoa to evade the host's immune response.

Sertoli cells also
1. nourish all spermatogenic cells

2. remove degenerating spermatogenic cells by phagocytosis

3. control release of spermatozoa into the lumen of the tubule

4. secrete the hormone **inhibin** which is a feedback inhibitor of gonadrotrophin release

5. secrete **androgen-binding protein** which concentrates testosterone in the seminiferous tubules where it promotes spermatogenesis

6. remove excess cytoplasm as spermatids develop into spermatozoa.

SPERMATOZOA

SECONDARY SPERMATOCYTES remain attached by cytoplasmic bridges which persist until sperm development is complete. This may be necessary since half the sperm carry an X chromosome and half a Y chromosome. The genes carried on the X chromosome and absent on the Y chromosome may be essential for survival of Y-bearing sperm, and hence for the production of male offspring.

SPERMATOGONIUM (Sperm mother cell) divides mitotically ($2n \begin{smallmatrix} \nearrow 2n \\ \searrow 2n \end{smallmatrix}$): the two products have different fates

– one remains to prevent depletion of the reservoir of these stem cells, the other loses contact with the basement membrane and becomes a primary spermatocyte.

PRIMARY SPERMATOCYTE has been formed from sperm mother cell and now undergoes a meiotic (reduction) division: the first meiotic division, which involves synapsis and offers the possibility of crossover, produces two secondary spermatocytes, and the second miotic division, which is effectively a modified mitosis, produces four spermatids i.e.

Primary $\xrightarrow[\text{I}]{\text{Meiosis}}$ Secondary
1 Spermatocyte   2 Spermatocytes
(diploid)   (haploid)

$\xrightarrow[\text{II}]{\text{Meiosis}}$ 4 Spermatids
(haploid)

SPERMATIDS become embedded in Sertoli cells and mature into spermatozoa, the process of **spermiogenesis**. This involves the development of an acrosome and a flagellum, and since no division is involved.

1 Spermatid $\longrightarrow$ 1 Spermatozoan
(haploid)   (haploid)

Structure of the ovary

reflects events of the menstrual cycle

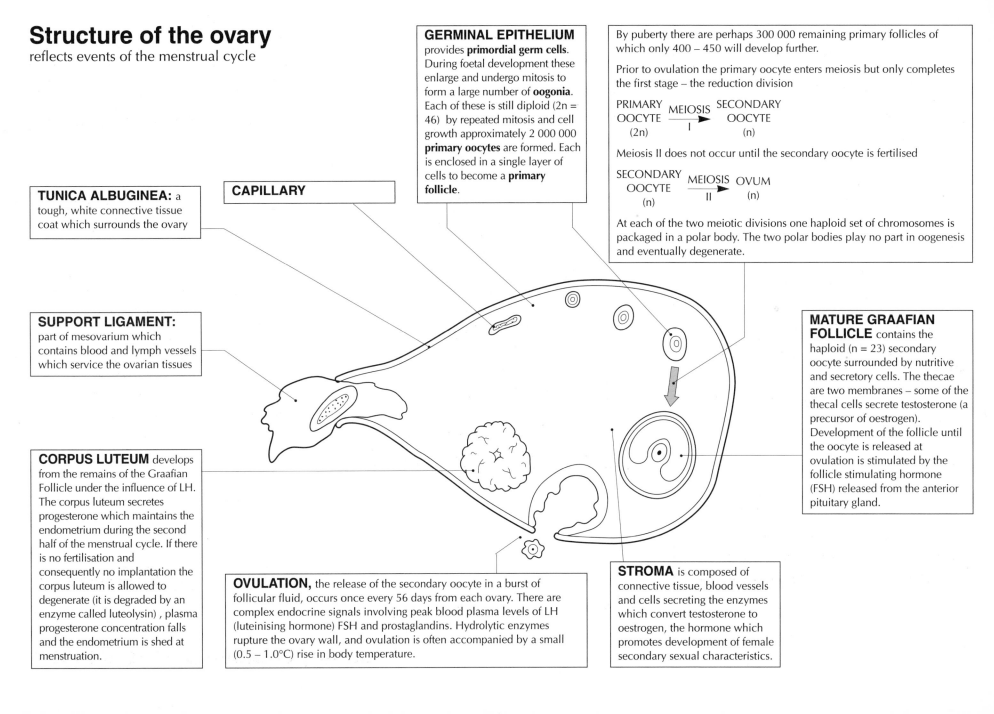

GERMINAL EPITHELIUM
provides **primordial germ cells**.
During foetal development these
enlarge and undergo mitosis to
form a large number of **oogonia**.
Each of these is still diploid (2n =
46) by repeated mitosis and cell
growth approximately 2 000 000
primary oocytes are formed. Each
is enclosed in a single layer of
cells to become a **primary
follicle**.

By puberty there are perhaps 300 000 remaining primary follicles of
which only 400 – 450 will develop further.

Prior to ovulation the primary oocyte enters meiosis but only completes
the first stage – the reduction division

PRIMARY   SECONDARY
OOCYTE $\xrightarrow[\text{I}]{\text{MEIOSIS}}$ OOCYTE
(2n)   (n)

Meiosis II does not occur until the secondary oocyte is fertilised

SECONDARY
OOCYTE $\xrightarrow[\text{II}]{\text{MEIOSIS}}$ OVUM
(n)   (n)

At each of the two meiotic divisions one haploid set of chromosomes is
packaged in a polar body. The two polar bodies play no part in oogenesis
and eventually degenerate.

CAPILLARY

TUNICA ALBUGINEA: a
tough, white connective tissue
coat which surrounds the ovary

SUPPORT LIGAMENT:
part of mesovarium which
contains blood and lymph vessels
which service the ovarian tissues

**MATURE GRAAFIAN
FOLLICLE** contains the
haploid (n = 23) secondary
oocyte surrounded by nutritive
and secretory cells. The thecae
are two membranes – some of the
thecal cells secrete testosterone (a
precursor of oestrogen).
Development of the follicle until
the oocyte is released at
ovulation is stimulated by the
follicle stimulating hormone
(FSH) released from the anterior
pituitary gland.

CORPUS LUTEUM develops
from the remains of the Graafian
Follicle under the influence of LH.
The corpus luteum secretes
progesterone which maintains the
endometrium during the second
half of the menstrual cycle. If there
is no fertilisation and
consequently no implantation the
corpus luteum is allowed to
degenerate (it is degraded by an
enzyme called luteolysin) , plasma
progesterone concentration falls
and the endometrium is shed at
menstruation.

OVULATION, the release of the secondary oocyte in a burst of
follicular fluid, occurs once every 56 days from each ovary. There are
complex endocrine signals involving peak blood plasma levels of LH
(luteinising hormone) FSH and prostaglandins. Hydrolytic enzymes
rupture the ovary wall, and ovulation is often accompanied by a small
(0.5 – 1.0°C) rise in body temperature.

STROMA is composed of
connective tissue, blood vessels
and cells secreting the enzymes
which convert testosterone to
oestrogen, the hormone which
promotes development of female
secondary sexual characteristics.

Human spermatozoon and oocyte

Acrosome is effectively an enclosed lysosome. It develops from the Golgi complex and contains hydrolytic enzymes – a hyaluronidase and several proteinases – which aid in the penetration of the granular layer and plasma membrane of the oocyte immediately prior to fertilization.

Nucleus contains the haploid number of chromosomes derived by meiosis from the male germinal cells – thus this genetic complement will be either an X or a Y heterosome plus 22 autosomes. Since the head delivers no cytoplasm, the male contributes no extranuclear genes or organelles.

Centriole: one of a pair, which lie at right angles to one another. One of the centrioles produces microtubules which elongate and run the entire length of the rest of the spermatozoon, forming the axial filament of the flagellum.

Mitochondria are arranged in a spiral surrounding the flagellum. They complete the aerobic stages of respiration to release the ATP required for contraction of the filaments, leading to 'beating' of the flagellum and movement of the spermatozoon.

PERIPHERAL MICROTUBULE

DYNEIN 'ARM' (acts as ATPase)

CENTRAL PAIR OF MICROTUBULES

Flagellum has the 9 + 2 arrangement of microtubules typical of such structures. The principal role of the flagella is to allow sperm to move close to the oocyte and to orientate themselves correctly prior to digestion of the oocytemembranes. The sperm are moved close to the oocyte by muscular contractions of the walls of the uterus and the oviduct.

HUMAN SPERMATOZOON (v.s.)

HEAD 5 μm

MID PIECE 7 μm

TAIL PIECE 45 μm

First polar body contains 23 chromosomes from the first meiotic division of the germ wall.

140 μm

HUMAN SECONDARY OOCYTE (v.s.)

Cumulus cells which once synthesized proteins and nucleic acids into the oocyte cytoplasm.

Zona pellucida will undergo structural changes at fertilization and form a barrier to the entry of more than one sperm.

23 chromosomes will complete second meiotic division on fertilization to provide *female haploid nucleus* (an ovum) and a second polar body.

Cytoplasm may contribute extranuclear genes and organelles to the zygote.

Cortical granules contain enzymes which are released at fertilization and alter the structure of the zona pellucida, preventing further sperm penetration, which would upset the just restored diploid number.

126 Human spermatozoon and oocyte

Events of the menstrual cycle

Follicle-stimulating hormone (FSH) initiates the development of several primary follicles (each containing a primary oocyte): one follicle continues to develop but the others degenerate by the process of follicular atresia. FSH also increases the activity of the enzymes responsible for formation of oestrogen. FSH, either synthetic or extracted from placentae, can increase fertility by promoting ovulation.

Oestrogen is produced by enzyme modification (in the stroma) of testosterone produced by the thecae of the developing follicle. Oestrogen has several effects:
1. It stimulates further growth of the follicle.
2. It promotes repair of the endometrium.
3. It acts as a feedback inhibitor of the secretion of FSH from the anterior pituitary gland.
4. From about day 11 it has a positive feedback action on the secretion of both LH and FSH.

Development of the follicle within the ovary is initiated by FSH but continued by LH. The Graafian follicle (Ⓐ) is mature by day 10–11 and ovulation occurs at day 14 (Ⓑ) following a surge of LH. The remains of the follicle become the corpus luteum (Ⓒ), which secretes steroid hormones. These steroid hormones inhibit LH secretion so that the corpus luteum degenerates and becomes the corpus albicans (Ⓓ).

The **endometrium** begins to thicken and become more vascular under the influence of the ovarian hormone oestrogen. Because of this thickening, to 4–6 mm, the time between menstruation and ovulation is sometimes called the **proliferative phase.**

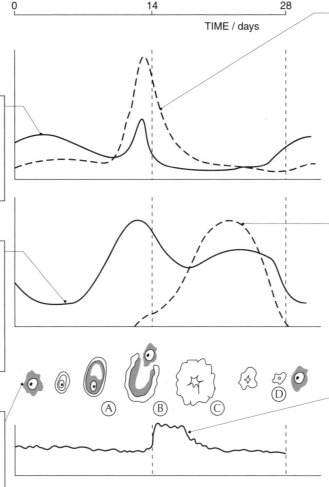

Luteinizing hormone (LH) triggers the secretion of testosterone by the thecae of the follicle, and when its concentration 'surges' it causes release of enzymes which rupture the wall of the ovary, allowing the secondary oocyte to be released at ovulation. After ovulation LH promotes development of the corpus luteum from the remains of the Graafian follicle.

Progesterone is secreted by the corpus luteum. It has several effects:
1. It prepares the endometrium for implantation of a fertilized egg by increasing vascularization, thickening and the storage of glycogen.
2. It begins to promote growth of the mammary glands.
3. It acts as a feedback inhibitor of FSH secretion, thus arresting development of any further follicles.
Oestrogen and progesterone combined act as contraceptive 'pills' by inhibition of follicle development and ovulation.

Body temperature rises by about 1°C at the time of ovulation. This 'heat' is used to determine the 'safe period' for the rhythm method of contraception.

During the post-ovulatory or luteal phase the **endometrium** becomes thicker with more tortuously coiled glands and greater vascularization of the surface layer, and retains more tissue fluid.

Menstruation is initiated by falling concentrations of oestrogen and progesterone as the corpus luteum degenerates.

At **menstruation** the *stratum functionalis* of the endometrium is shed, leaving the *stratum basilis* to begin proliferation of a new *functionalis*.

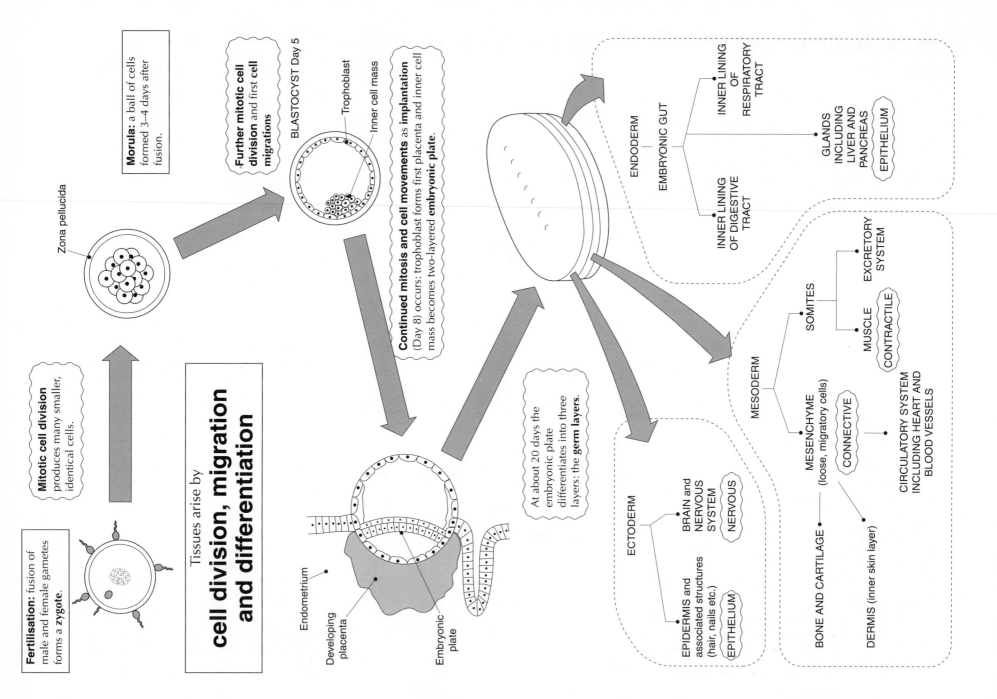

Fertilisation: fusion of male and female gametes forms a **zygote.**

Mitotic cell division produces many smaller, identical cells.

Tissues arise by

cell division, migration and differentiation

Zona pellucida

Morula: a ball of cells formed 3–4 days after fusion.

Further mitotic cell division and first **cell migrations**

BLASTOCYST Day 5

Trophoblast

Inner cell mass

Continued mitosis and cell movements as **implantation** (Day 8) occurs: trophoblast forms first placenta and inner cell mass becomes two-layered **embryonic plate.**

Endometrium

Developing placenta

Embryonic plate

At about 20 days the embryonic plate differentiates into three layers: the **germ layers.**

ENDODERM

EMBRYONIC GUT

INNER LINING OF DIGESTIVE TRACT

INNER LINING OF RESPIRATORY TRACT

GLANDS INCLUDING LIVER AND PANCREAS

EPITHELIUM

MESODERM

SOMITES

EXCRETORY SYSTEM

MUSCLE — CONTRACTILE

MESENCHYME (loose, migratory cells)

CONNECTIVE

CIRCULATORY SYSTEM INCLUDING HEART AND BLOOD VESSELS

BONE AND CARTILAGE

DERMIS (inner skin layer)

ECTODERM

BRAIN AND NERVOUS SYSTEM — NERVOUS

EPIDERMIS and associated structures (hair, nails etc.) — EPITHELIUM

128 Cell division, migration, and differentiation

Functions of the placenta

Barrier: the placenta limits the transfer of solutes and blood components from maternal to fetal circulation. Cells of the maternal immune system do not cross – this minimizes the possibility of immune rejection (although antibodies may cross which may cause haemolysis of fetal blood cells if **Rhesus** antibodies are present in the maternal circulation). The placenta is **not** a barrier to heavy metals such as lead, to nicotine, to HIV and other viruses, to heroin and to toxins such as DDT. Thus the **Rubella** (German measles) virus may cross and cause severe damage to eyes, ears, heart and brain of the fetus, the sedative Thalidomide caused severely abnormal limb development, and some children are born already addicted to heroin or HIV positive.

Immune protection: protective molecules (possibly including HCG) cover the surface of the early placenta and 'camouflage' the embryo which, with its complement of paternal genes, might be identified by the maternal immune system and rejected as tissue of 'non-self' origin.

Site of exchange of many solutes between maternal and fetal circulations. Oxygen transfer to the fetus is aided by fetal haemoglobin with its high oxygen affinity, and soluble nutrients such as glucose and amino acids are selectively transported by membrane-bound carrier proteins. Carbon dioxide and urea diffuse from fetus to mother along diffusion gradients maintained over larger areas of the placenta by countercurrent flow of fetal and maternal blood. In the later stages of pregnancy antibodies pass from mother to fetus – these confer immunity in the young infant, particularly to some gastro-intestinal infections.

Endocrine function: cells of the chorion secrete a number of hormones:
1. **HCG (human chorionic gonadotrophin)** maintains the corpus luteum so that this body may continue to secrete the progesterone necessary to continue the development of the endometrium. HCG is principally effective during the first 3 months of pregnancy and the overflow of this hormone into the urine is used in diagnosis of the pregnant condition.
2. **Oestrogen** and **progesterone** are secreted as the production of these hormones from the degenerating corpus luteum diminishes.
3. **Human placental lactogen** promotes milk production in the mammary glands as birth approaches.
4. **Prostaglandins** are released under the influence of fetal adrenal steroid hormones. These prostaglandins are powerful stimulators of contractions of the smooth muscle of the uterus, the contractions which constitute labour and eventually expel the fetus from the uterus.

After expulsion of the fetus, further contractions of the uterus cause detachment of the placenta (spontaneous constriction of uterine artery and vein limit blood loss) – the placenta, once delivered, is referred to as the **afterbirth.** The placenta may be used as a source of hormones (e.g. in 'fertility pills'), as tissue for burn grafts and to supply veins for blood vessel grafts.

Hormones and parturition

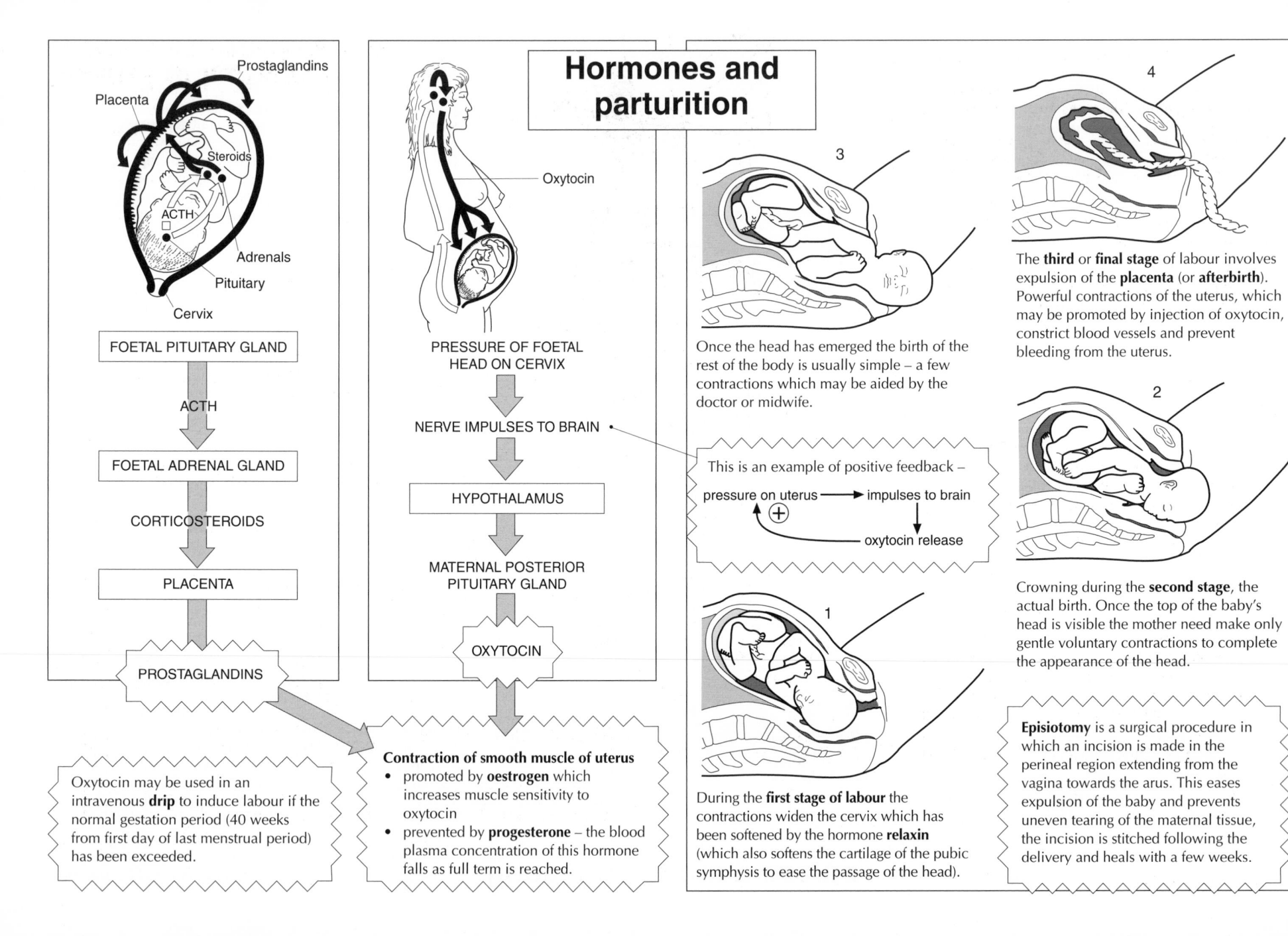

FOETAL PITUITARY GLAND

↓ ACTH

FOETAL ADRENAL GLAND

↓ CORTICOSTEROIDS

PLACENTA

↓

PROSTAGLANDINS

Oxytocin may be used in an intravenous **drip** to induce labour if the normal gestation period (40 weeks from first day of last menstrual period) has been exceeded.

PRESSURE OF FOETAL HEAD ON CERVIX

↓

NERVE IMPULSES TO BRAIN

↓

HYPOTHALAMUS

↓

MATERNAL POSTERIOR PITUITARY GLAND

↓

OXYTOCIN

Contraction of smooth muscle of uterus
- promoted by **oestrogen** which increases muscle sensitivity to oxytocin
- prevented by **progesterone** – the blood plasma concentration of this hormone falls as full term is reached.

This is an example of positive feedback –

pressure on uterus ⟶ impulses to brain
⊕ ↑ ↓
oxytocin release

Once the head has emerged the birth of the rest of the body is usually simple – a few contractions which may be aided by the doctor or midwife.

During the **first stage of labour** the contractions widen the cervix which has been softened by the hormone **relaxin** (which also softens the cartilage of the pubic symphysis to ease the passage of the head).

The **third** or **final stage** of labour involves expulsion of the **placenta** (or **afterbirth**). Powerful contractions of the uterus, which may be promoted by injection of oxytocin, constrict blood vessels and prevent bleeding from the uterus.

Crowning during the **second stage**, the actual birth. Once the top of the baby's head is visible the mother need make only gentle voluntary contractions to complete the appearance of the head.

Episiotomy is a surgical procedure in which an incision is made in the perineal region extending from the vagina towards the arus. This eases expulsion of the baby and prevents uneven tearing of the maternal tissue, the incision is stitched following the delivery and heals with a few weeks.

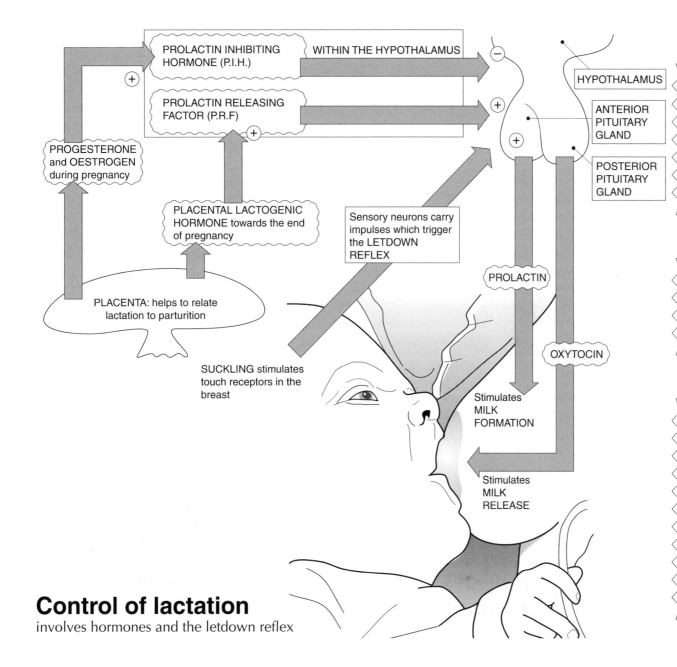

PROLACTIN INHIBITING HORMONE (P.I.H.)

PROLACTIN RELEASING FACTOR (P.R.F)

WITHIN THE HYPOTHALAMUS

HYPOTHALAMUS

ANTERIOR PITUITARY GLAND

POSTERIOR PITUITARY GLAND

PROGESTERONE and OESTROGEN during pregnancy

PLACENTAL LACTOGENIC HORMONE towards the end of pregnancy

PLACENTA: helps to relate lactation to parturition

Sensory neurons carry impulses which trigger the LETDOWN REFLEX

PROLACTIN

OXYTOCIN

SUCKLING stimulates touch receptors in the breast

Stimulates MILK FORMATION

Stimulates MILK RELEASE

PROLACTIN AND CONTRACEPTION
Prolactin interferes with the action of **luteinising hormone releasing factor**, thus inhibiting the secretion of **luteinising hormone** (LH) and **follicle stimulating hormone** (FSH). As a result ovulation and the normal menstrual cycle ceases – regular and frequent suckling may act as a natural contraceptive in some women.

GETTING IN SHAPE Breastfeeding stimulates continued release of **oxytocin**. This helps the uterus to recover from childbirth by stimulating it to return to non-pregnant size.

BREAST IS BEST!
- for the first few days the mammary glands secrete **colostrum** – contains no fat and little lactose but delivers antibodies which confer **passive immunity**.

- true milk contains a balanced mixture of nutrients, it has **less** protein, sodium and calcium than cow's milk, but **more** vitamin C and vitamin D.

- immunoglobulins, initially IgG and later IgA, are important in offering protection against microbes which might colonise the baby's gut and cause disease

- it's also the **right temperature**, **cheap** and **available on demand!**

Control of lactation
involves hormones and the letdown reflex

The foetus

Heart rate and **stroke volume increase** to raise cardiac output by 30–35%.

Blood volume and red cell count increase – the additional 1 dm³ of blood actually **decreases** RBC concentration, causing **physiological anaemia of pregnancy**.

Thoracic cavity widens to compensate for raised position of diaphragm so that tidal volume and pulmonary ventilation of maternal lungs increases. Together with greater cardiac output this ensures delivery of extra oxygenated blood to the foetus.

Kidney blood flow and renal filtration rate increase – fewer solutes are reabsorbed which may cause **glycosuria** (glucose excretion) during pregnancy.

Tissue fluid is retained, especially in the lower limbs, as the increased blood volume dilutes plasma proteins: the reduced solute potential of the plasma thus limits tissue fluid reabsorption into the blood.

Parathormone secretion increases causing release of calcium and phosphate ions from the maternal skeleton – these are transferred to the foetus for bone growth.

Breasts increase in size as mammary glands and lactiferous ducts develop during late pregnancy.

Adipose tissue deposition increases – provides a source of energy for late pregnancy and lactation.

Uterus expands to accommodate the developing foetus – growth of muscle fibres and connective tissue is stimulated by **oestrogen**.

9 months
8
6
4
3
Contours of uterus during pregnancy

Blood flow to limbs increases to permit loss of excess heat generated by increased metabolism during foetal development – limbs and face often become pink due to vasodilation of subcutaneous blood vessels.

CHANGES DURING PREGNANCY PROMOTE DEVELOPMENT OF THE FOETUS

DIFFERENCES BETWEEN FOETAL AND ADULT CIRCULATION result from the non-use of the lungs, the kidneys and the gut in the foetus.

Ductus arteriosus allows oxygenated blood from the **right** ventricle to bypass the lungs and pass via the aorta into the systemic circulation

Foramen ovale allows **oxygenated blood from the placenta** to pass directly to the left atrium

Ductus venosus carries oxygenated blood from the placenta to the inferior vena cava – vena cava thus carries oxygenated blood

Placenta – performs functions of lungs, kidneys and gut

Umbilical vein – blood with **high** O₂/nutrient, but low CO₂/urea, content

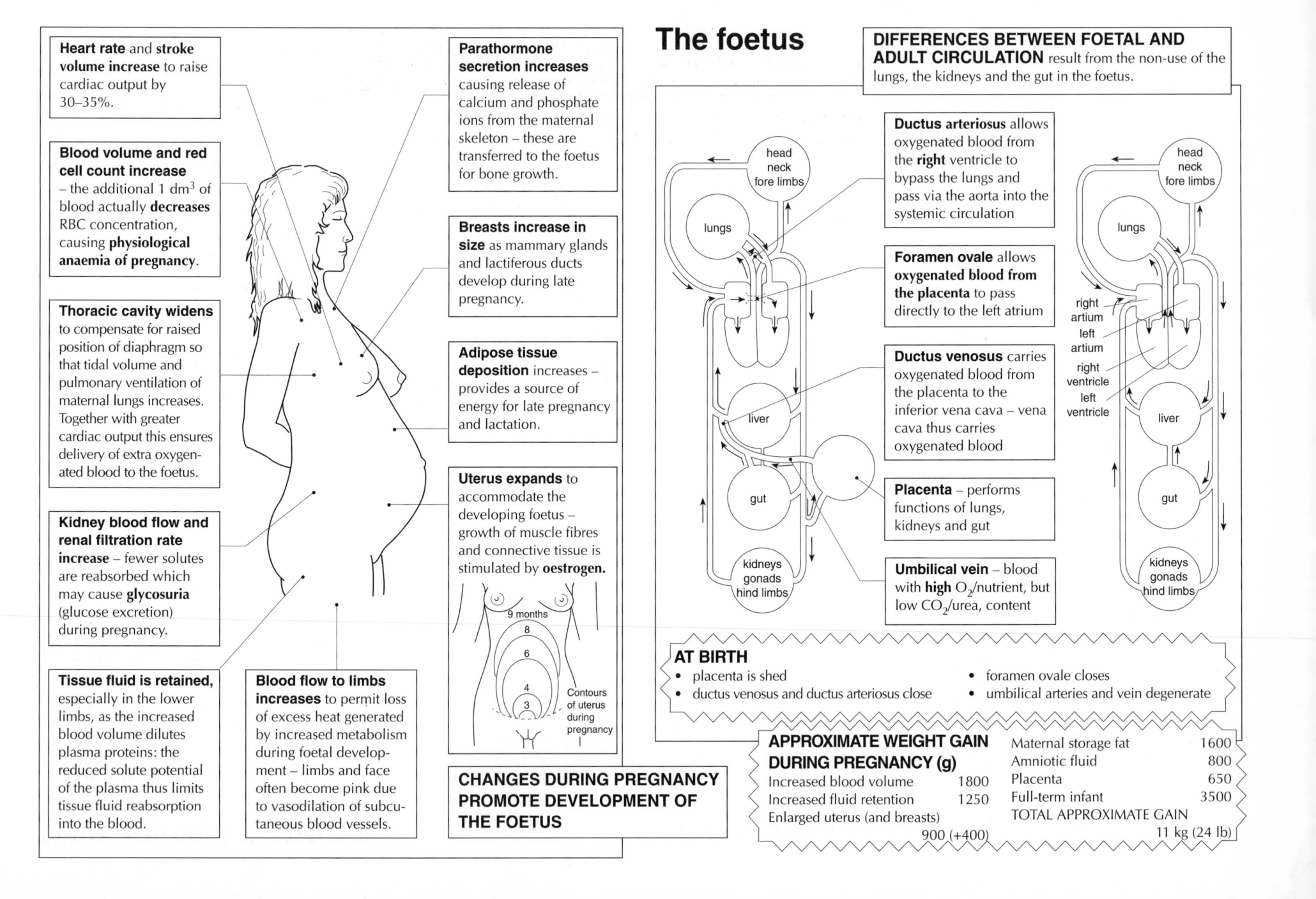

head neck fore limbs · lungs · liver · gut · kidneys gonads hind limbs

head neck fore limbs · lungs · right artium · left artium · right ventricle · left ventricle · liver · gut · kidneys gonads hind limbs

AT BIRTH
- placenta is shed
- ductus venosus and ductus arteriosus close
- foramen ovale closes
- umbilical arteries and vein degenerate

APPROXIMATE WEIGHT GAIN DURING PREGNANCY (g)

Increased blood volume	1800	
Increased fluid retention	1250	
Enlarged uterus (and breasts)		
	900 (+400)	
Maternal storage fat	1600	
Amniotic fluid	800	
Placenta	650	
Full-term infant	3500	
TOTAL APPROXIMATE GAIN	11 kg (24 lb)	

Measurement of growth

METHODS OF GROWTH MEASUREMENT – all involve measuring some parameter of the individual and then plotting the results against time

- take **samples** from populations wherever possible, and calculate means to minimise the effect of any individual result

- choose the correct parameter

Supine length – eliminates the problem of spinal compression or poor muscular tone. Most useful directly comparative linear measurement

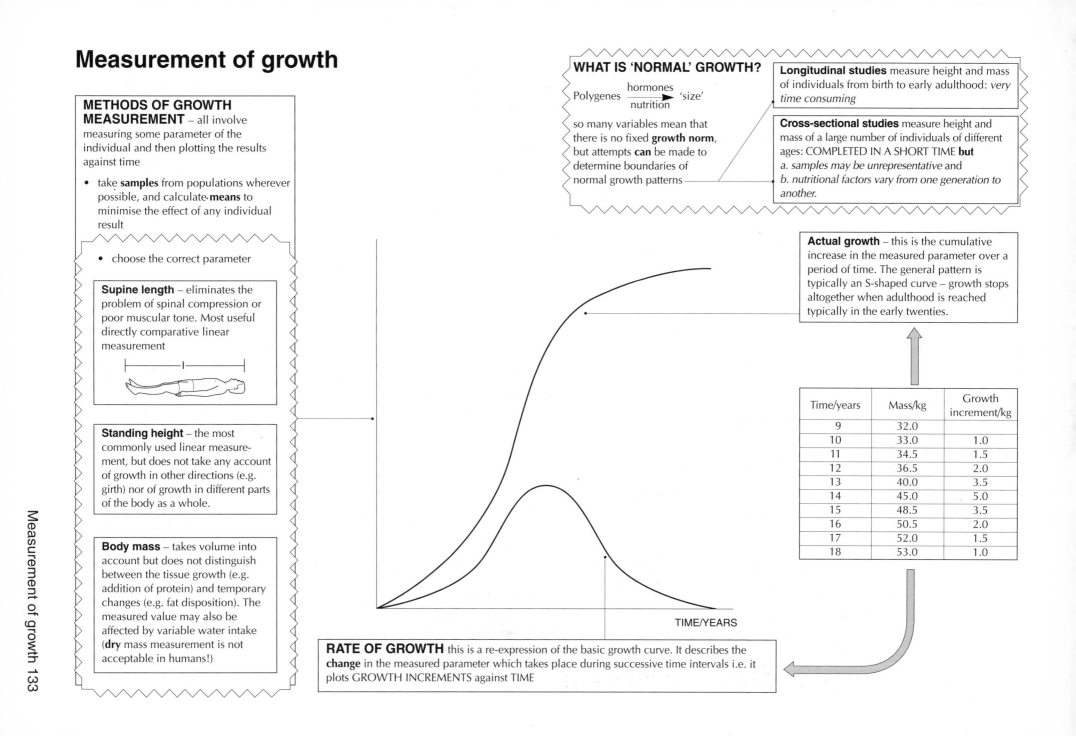

Standing height – the most commonly used linear measurement, but does not take any account of growth in other directions (e.g. girth) nor of growth in different parts of the body as a whole.

Body mass – takes volume into account but does not distinguish between the tissue growth (e.g. addition of protein) and temporary changes (e.g. fat disposition). The measured value may also be affected by variable water intake (**dry** mass measurement is not acceptable in humans!)

WHAT IS 'NORMAL' GROWTH?

Polygenes $\xrightarrow[\text{nutrition}]{\text{hormones}}$ 'size'

so many variables mean that there is no fixed **growth norm**, but attempts **can** be made to determine boundaries of normal growth patterns

Longitudinal studies measure height and mass of individuals from birth to early adulthood: *very time consuming*

Cross-sectional studies measure height and mass of a large number of individuals of different ages: COMPLETED IN A SHORT TIME **but**
a. *samples may be unrepresentative* and
b. *nutritional factors vary from one generation to another.*

Actual growth – this is the cumulative increase in the measured parameter over a period of time. The general pattern is typically an S-shaped curve – growth stops altogether when adulthood is reached typically in the early twenties.

Time/years	Mass/kg	Growth increment/kg
9	32.0	
10	33.0	1.0
11	34.5	1.5
12	36.5	2.0
13	40.0	3.5
14	45.0	5.0
15	48.5	3.5
16	50.5	2.0
17	52.0	1.5
18	53.0	1.0

TIME/YEARS

RATE OF GROWTH this is a re-expression of the basic growth curve. It describes the **change** in the measured parameter which takes place during successive time intervals i.e. it plots GROWTH INCREMENTS against TIME

Growth patterns in humans

Changes in FORM (the **qualitative** changes alongside growth) are caused by two factors

Timing: when do organs commence growth

Rate: organs may grow at different rates from one another and from the body as a whole = ALLOMETRIC GROWTH

LYMPHOID TISSUE: this tissue is found throughout the body in lymph nodes, and aggregated in tonsils, adenoids and thymus gland. A major function is the production of antibodies and the rapid growth up to the early teens corresponds to the child's rapid acquisition of immunity to a considerable number of infectious conditions which may be encountered for the first time.

BRAIN: at birth the brain is already 25% of the eventual adult size, by one year it is 75% of the adult size and by 7–8 years of age brain growth is almost complete. At birth the brain contains its full complement of neurones so that brain growth after birth is due to increase in

- size (length) of individual cells
- number of subsidiary fibres (e.g. dendrons) as cross-connections are formed between cells
- development of the insulating myelin sheath around the neurones.

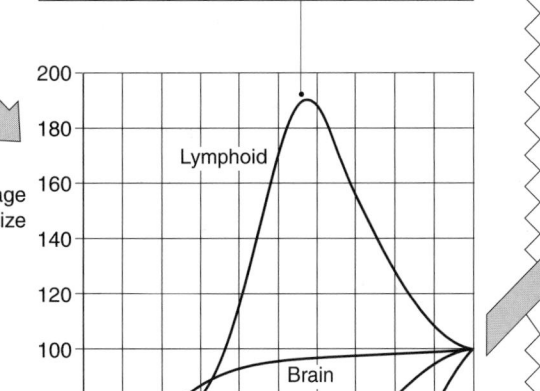

Size as percentage of final size

REPRODUCTIVE ORGANS: both **primary** (ovary, testis) and **secondary** (vulva and breasts, penis) sexual organs show little change until puberty. At this time PITUITARY GONADOTROPHINS followed by SEX STEROIDS (oestrogen, testosterone) initiate rapid growth which continues through adolescence.

THE GENERAL GROWTH CURVE does not indicate

a. **Changes in proportion:** these occur because of the different rates of growth/timing of growth for different parts of the body ...

- the proportion of body length occupied by the head falls from about 25% to about 15% – a young child cannot reach over its head to touch the opposite shoulder
- the legs increase in relative length – from about 35% in a 2-year old to about 50% in a 16-year old

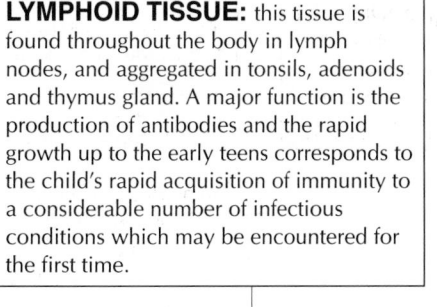

Newborn   2 years   7 years   18 years

b. **Effects of growth spurts:** when these occur ('infant' between 0–2, 'adolescent' between 14–17) the body length increases more rapidly than the body mass, giving a 'tall and gangly' appearance. The increase in height is followed by a period of 'filling out' when body mass (much of which is muscle) increases. Growth spurts occur at different ages in boys and girls.

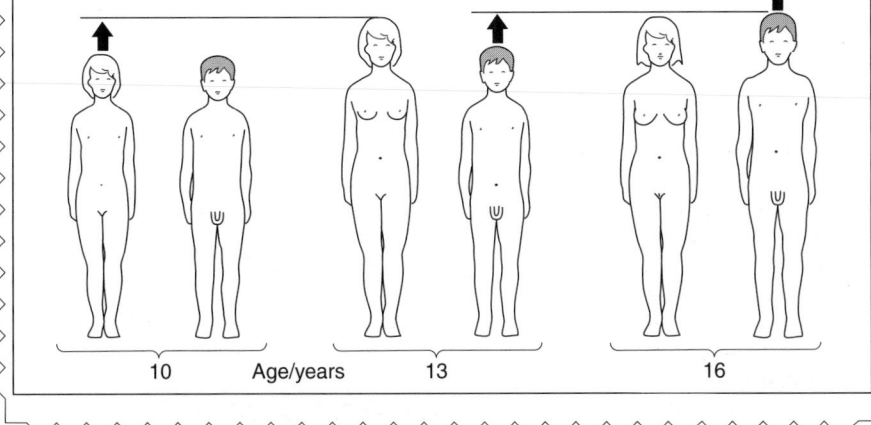

10   Age/years   13   16

Ageing

involves **age-related** (i.e. programmed from birth) and **age-associated** (accumulated throughout life) changes

ABILITIES: age-related changes in the **nervous system** influence aspects of **perception**, **intellect** and **activity**

vision: deterioration occurs due to **lens hardening and opacity** and damage to **retina** as blood vessels infiltrate

hearing problems are largely caused by **nerve cell degeneration**, causing deafness to sounds of high frequency. There may also be consequent difficulties in localisation of sound sources.

changes rather than **deteriorates**
- learning continues so vocabulary increases.
- short-term memory deteriorates
- ability to learn ideas is retained, but information processing may be slower

sleep patterns as pattern of circadian rhythms changes
- less time spent in deep sleep (less sleep-walking and nightmares)
- reduced overall sleeping time (8h. per day at age 18, 7h. per day at 60)

METABOLISM

- reduced **appetite** as number of taste buds and olfactory neurones decreases
- less efficient **digestion** – juice secretion diminishes, e.g. less acid in stomach-contributes to reduced Ca^{2+} absorption
- less efficient **temperature regulation** (poor vasoconstriction, reduced sweating/shivering ability) → danger of **hypothermia**
- deterioration in **immune system** includes **decrease** in antibody production, T cell responses and interleukin secretion but **increase** in production of autoantibodies. As a result the elderly are **susceptible to infection**
- **liver infection** declines, so that the clearance of toxins is reduced
- **kidney function** deteriorates as the number of nephrons and their sensitivity to endocrine control declines. Thus
  - (a) reduced filtration rate means slower removal of water-soluble toxins
  - (b) reducing concentrating power means greater susceptibility to fluid deprivation

Overall, maintenance of fluid, electrolyte and acid-base balance is more difficult in the elderly.

APPEARANCE: changes are due to **skin, hair** and **nails** and to **body build**

- loss of height as vertebrae and discs compress
- 'stooping' as ligaments and muscles become less supportive
- 'fatness' if energy intake is not matched to reduced metabolism and less vigorous exercise

- skin thinner and more wrinkled due to reduced elasticity of connective tissues and loss of subcutaneous fat
- hair 'greys' as melanocytes are lost and 'thins' as number of follicles declines. Hair loss may be androgen-related in males.

AGE–RELATED FACIAL CHANGES

18 years   30 years   50 years   70 years

PHYSICAL FITNESS AND STRENGTH

Decline due to changes **muscle mass, circulatory** and **respiratory efficiency**

diminishes as change in lifestyle (less exercise) and less efficient metabolsim (less efficient nutrient supply) combine with loss of motor neurones

hardening (**arteriosclerosis**) and narrowing (**atherosclerosis**) of arteries, together with reduced elasticity of cardiac muscle and decline in pacemaker cells may reduce circulatory efficiency by 30% at age 60

decreases in **vital capacity** and **expiratory force** mean that oxygen uptake may fall to 50% of that of an 18 year old

SEXUAL AND REPRODUCTIVE ACTIVITY: changes are associated with alterations in
- urinogenital system • hypothalmic and pituitary secretions • hormone-sensitivity of reproductive tissues

In **females** these changes culminate in **menopause** (age range 45–55 years) which marks the end of reproductive, though not sexual, activity. In **males** there is no equivalent to the menopause and reproductive capability may continue, although diminished until 90-plus years of age.

Two features of menopause are **hot flushes** and progressive **osteoporosis** – both may be diminished by **hormone replacement therapy** (HRT) in which oestrogen levels are maintained artificially.

Fibrous capsule is formed of a connective tissue with an abundance of collagen fibres. It is attached to the periosteum of the articulating bones, and may be almost indistinguish-able from extracapsular accessory ligament. The flexibility of the fibrous capsule permits movement at a joint and the great tensile strength of the collagen fibres resists dislocation.

Synovial membrane is internal to the fibrous capsule and is composed of loose connective tissue with an abundance of elastin fibres and variable amounts of adipose tissue. The membrane secretes *synovial fluid* which lubricates the joint and nourishes the articular cartilage.

Intracapsular ligament is located inside the articular capsule, but surrounded by folding of the synovial membrane. Examples are the *cruciate ligaments* of the knee joint.

Meniscal cartilage (mediscus, articular disc) is a pad of fibrocartilage which lies between the articular surfaces of the bones. By modifying the shapes of the articulating surfaces they allow two bones of different shapes to fit tightly together. Menisci also help to maintain the stability of the joint, and direct the flow of synovial fluid to areas of greatest friction.

Synovial fluid consists of hyaluronic acid and interstitial fluid formed from blood plasma. It is similar in appearance and consistency to egg white – in joints with little or no movement the fluid is viscous, but becomes less so as movement increases. The fluid also contains phagocytic cells which remove microbes and debris that results from wear and tear in the joint.

Osteoarthritis is a degenerative joint disease in which articular cartilage is destroyed and replaced by bony lumps which restrict joint movement and cause pain. Exercise seems to reduce development of this condition because exercise promotes synovial fluid formation, and this fluid provides nutrients for the cartilage. Osteoarthritis is most common in weight-bearing joints and may be accelerated by a blow (e.g. kick) to the joint.

Articular cartilage of synovial joints is hyaline cartilage. It is able to reduce friction in the joint since it is coated by synovial fluid – as the load on the joint increases the spongy nature of the cartilage allows it to absorb water and so increase the proportion of the lubricating hyaluronic acid in the lubricating film. The cartilage may also function as a shock absorber although much of the 'impact loading' on a joint is probably absorbed by the thicker trabecular (honey-combed) bone which lies beneath it.

Extracapsular ligament is one of the possible *accessory ligaments* which offer additional stability to a joint. 'Extracapsular' means outside the articular capsule, although often the ligament is fused with the capsule. An example is the *fibular collateral ligament* of the knee joint.

Bursae are sac-like structures with walls of connective tissue lined by synovial membrane, and filled with a fluid similar in composition and function to synovial fluid. They are found between muscle and bone, tendon and bone and ligament and bone, and reduce the friction between one structure and another as movement takes place.

Synovial joints have a space between the articulating bones and are freely movable.

THE MOVEMENTS POSSIBLE AT SYNOVIAL JOINTS ARE:

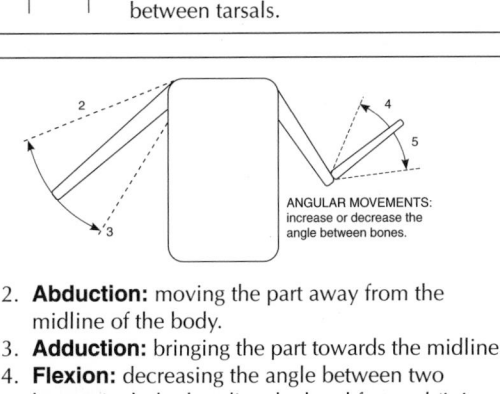

1. **Gliding:** one part slides upon another without any angular or rotary motion, e.g. joints between carpals and between tarsals.

ANGULAR MOVEMENTS: increase or decrease the angle between bones.

2. **Abduction:** moving the part away from the midline of the body.
3. **Adduction:** bringing the part towards the midline.
4. **Flexion:** decreasing the angle between two bones, includes bending the head forward (joint between the occipital bone and the atlas).
5. **Extension:** increasing the angle between two bones, includes returning the head to the anatomical position.

6. and 7. **Rotation:** turning upon an axis.
8. **Circumduction:** moving the extremity of the part around a circle so that the whole part describes a cone.

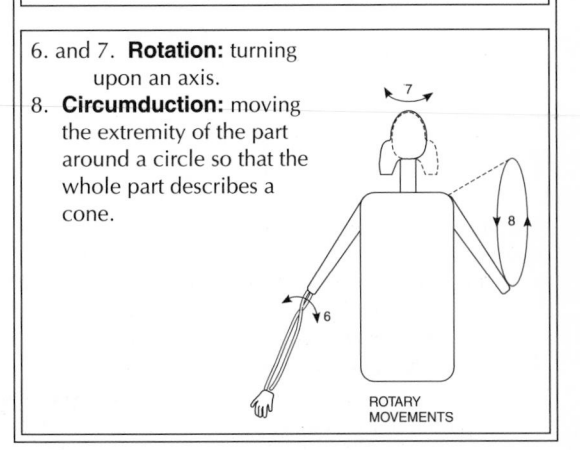

ROTARY MOVEMENTS

Humans are highly evolved primates

TYPICAL PRIMATES SHARE CERTAIN CHARACTERISTICS
- opposable thumb (can be held opposite the fingers)
- prehensile (grasping) hands and feet
- fingernails replace claws
- single pair of mammary glands
- 32 teeth in permanent, heterodont dentition
- flattened face offers stereoscopic vision

MAJOR STEPS IN HUMAN EVOLUTION
- larger brain
- facial changes
- bipedal gait
- more nimble hands
- better forms of communication
- use and production of tools
- co-operation in food production, defence, building and care of young

APE ANATOMY
- dense hair covers most of skin
- long arms, narrow pelvis, bent knees and fat feet all contribute to shuffling quadrupedal gait
- skull hangs forward from vertebral column
- face larger than cranium – cranial volume only 50% of that of human
- flattened nose, prominent eye ridges, thin lips and massive jaw – eating more significant than communication.

HUMAN ANATOMY
- very limited area of body surface is covered by hair
- legs longer than arms, wide pelvis and the ability to straighten the knee all contribute to **bipedal gait**
- arched feet/large buttocks mean more efficient walking
- skull supported on top of vertebral column
- cranium larger than face – large volume for brain with more extensive cerebrum
- protruding nose, flattened jaws, small eyebrow ridges but large lips contribute to wide range of facial expression.

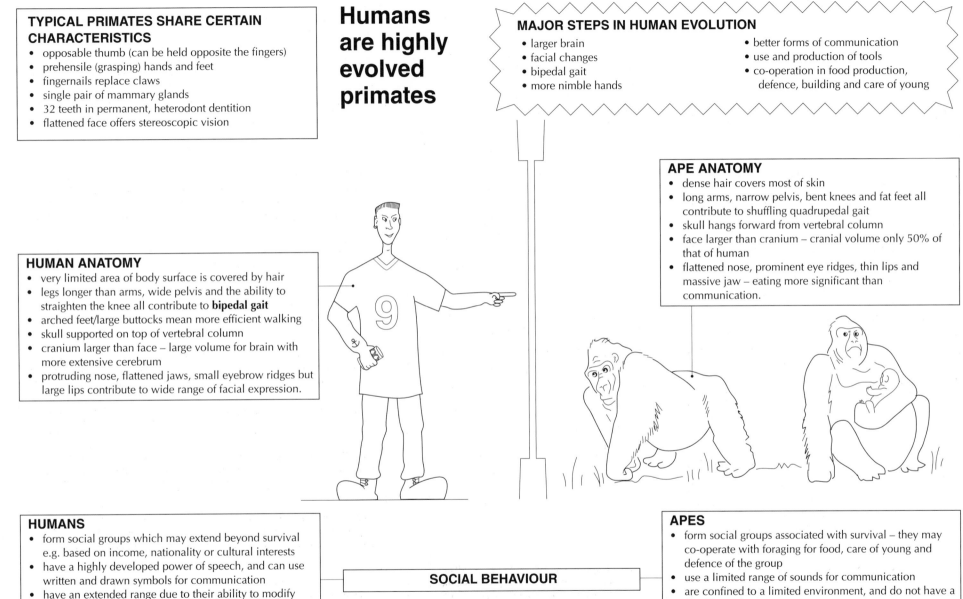

HUMANS
- form social groups which may extend beyond survival e.g. based on income, nationality or cultural interests
- have a highly developed power of speech, and can use written and drawn symbols for communication
- have an extended range due to their ability to modify their environment – make homes in fixed position
- make complex tools, and have learned to control fire for heating and cooking.

SOCIAL BEHAVIOUR

APES
- form social groups associated with survival – they may co-operate with foraging for food, care of young and defence of the group
- use a limited range of sounds for communication
- are confined to a limited environment, and do not have a fixed home base
- do not make tools, but may use objects as tools to perform simple tasks.

Human evolution

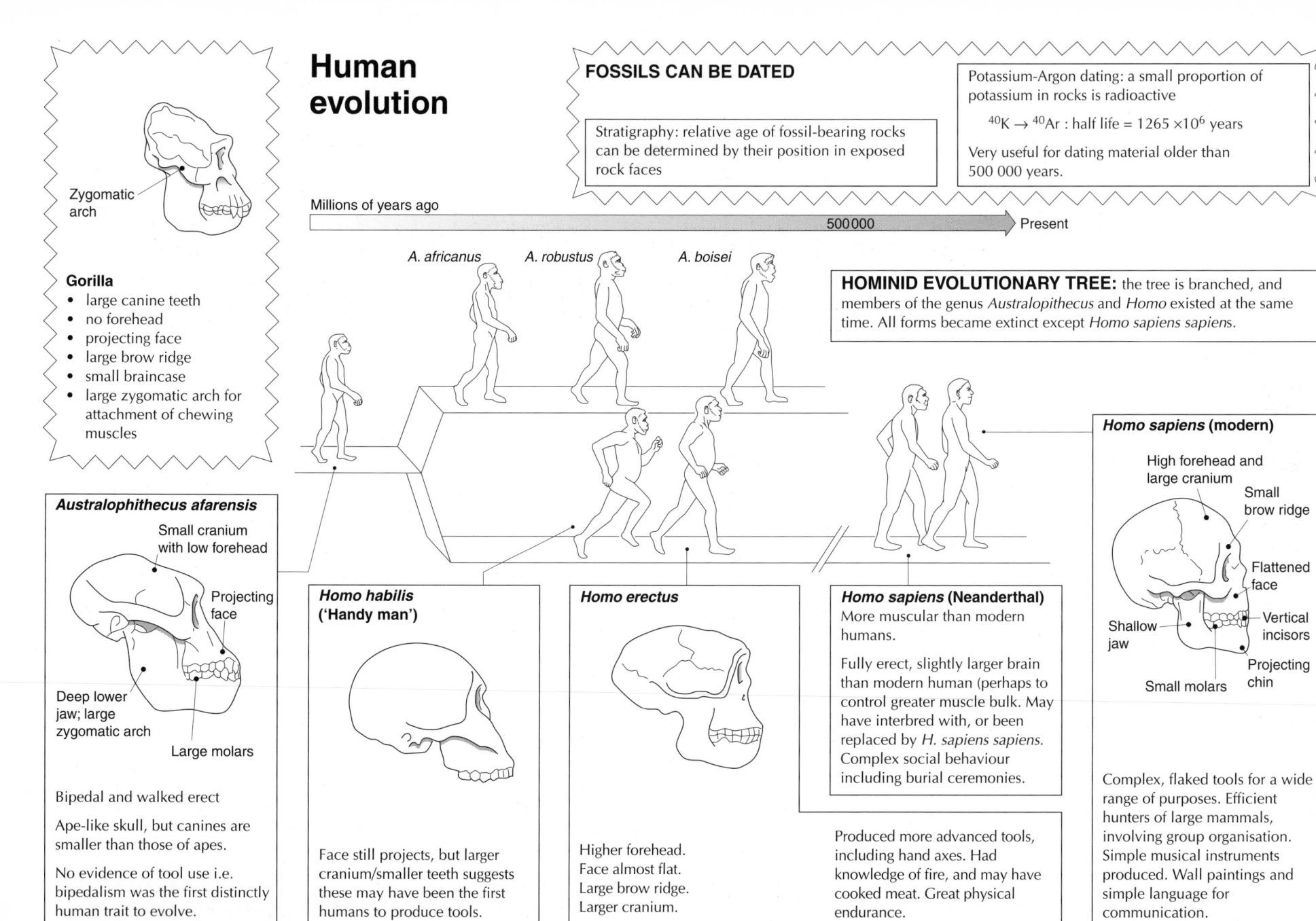

Zygomatic arch

Gorilla
- large canine teeth
- no forehead
- projecting face
- large brow ridge
- small braincase
- large zygomatic arch for attachment of chewing muscles

FOSSILS CAN BE DATED

Stratigraphy: relative age of fossil-bearing rocks can be determined by their position in exposed rock faces

Potassium-Argon dating: a small proportion of potassium in rocks is radioactive

$^{40}K \rightarrow {}^{40}Ar$: half life = 1265×10^6 years

Very useful for dating material older than 500 000 years.

Millions of years ago

500 000   Present

A. africanus   A. robustus   A. boisei

HOMINID EVOLUTIONARY TREE: the tree is branched, and members of the genus *Australopithecus* and *Homo* existed at the same time. All forms became extinct except *Homo sapiens sapiens*.

Australophithecus afarensis

Small cranium with low forehead

Projecting face

Deep lower jaw; large zygomatic arch

Large molars

Bipedal and walked erect

Ape-like skull, but canines are smaller than those of apes.

No evidence of tool use i.e. bipedalism was the first distinctly human trait to evolve.

Homo habilis ('Handy man')

Face still projects, but larger cranium/smaller teeth suggests these may have been the first humans to produce tools.

Homo erectus

Higher forehead. Face almost flat. Large brow ridge. Larger cranium.

Homo sapiens (Neanderthal)
More muscular than modern humans.

Fully erect, slightly larger brain than modern human (perhaps to control greater muscle bulk. May have interbred with, or been replaced by *H. sapiens sapiens*. Complex social behaviour including burial ceremonies.

Produced more advanced tools, including hand axes. Had knowledge of fire, and may have cooked meat. Great physical endurance.

Homo sapiens (modern)

High forehead and large cranium

Small brow ridge

Flattened face

Shallow jaw

Vertical incisors

Projecting chin

Small molars

Complex, flaked tools for a wide range of purposes. Efficient hunters of large mammals, involving group organisation. Simple musical instruments produced. Wall paintings and simple language for communication.

INDEX

colon 66
colostrum 76, 131
combustion 62, 64, 65
community 56
competition 58
complement 94, 95
concentration gradient 12
condensation 15, 18
condenser lens 6, 7
conditioned reflex 70
cone 115, 117
conformation of protein 18, 22
conjunctiva 114
consumer 60
cornea 114
corpus luteum 125, 127
cortex, of kidney 107
cortex, visual 115
countercurrent multiplication 107
covalent bond 14, 18
cramp 123
crossing-over 43, 48
cross-sectional growth study 133
cultural evolution 137
curare 113
cyanide 22
cyclic AMP 16, 119
cystic fibrosis 39, 44
cytochrome 79
cytochrome oxidase 22
cytokinesis 42, 43
cytokinin 94
cytoplasm 5, 8, 27
cytosine 32

D

deamination 106
decomposers 62, 63
deforestation 65
deletion 48
dendrite 109
denitrification 63
deoxyribose 16
depolarisation 110, 112
detoxification 106
development 76
diabetes 104
diaorrhoea 16
diaphragm 66, 77, 78, 81
diet, adequate and balanced 71
diet, crash 74
diet, weight control 74
dieting 74
diffusion 11, 12
diffusion, facilitated 11
digestion 66, 67, 69
dihybrid 45
disaccharide 16
disease 99
disulphide bridge 18
DIT (dietary induced thermogenesis) 75
DNA (deoxyribonucleic acid) 5, 8, 9,
    29, 32, 33, 34, 37, 38, 40, 41
DNA fingerprint 38
DNA profile 38
dorsal root ganglion 120
double helix 32

doubling time 58
Down's syndrome 48
DPG (diphosphoglycerate) 84
ductus arteriosus 132
ductus venosus 132
duodenum 66

E

ECG (electrocardiogram) 89
ecosystem 56, 60
ectoderm 128
electron beam 6, 7
electron carrier 25, 26, 27
electron transport chain 25, 26, 27, 30
electrophoresis 38
embolism 91
embryo 128
emphysema 77, 79, 80
emulsification 67
emulsion test 14, 17
endocrine organ 118
endoderm 128
endometrium 127
endonuclease 37
endoplasmic reticulum 8
endothermy 102, 103
energy 28, 75
environmental resistance 58
enzyme 20, 21, 22, 23
enzyme immobilisation 23
enzyme-substrate complex 20
eosinophil 83
epiglottis 66, 77
episiotomy 130
epithelium, ciliated 77
equator 42, 43
erythrocyte 83, 86
eukaryote 9, 40
exercise 90, 92, 123
expiration 78, 81
expiratory reserve volume 80
eye 114

F

factor 8 37
FAD (flavine adenine dinucleotide) 26
fat 17, 19, 67, 104, 106
fatigue 123
fatty acid 14, 21, 67, 71
fermentation 64
fermenter 37, 101
ferritin 106
fertilisation 128
fever 102
fibre 71
fibrin 18, 83
Fick's Law 12
flagellum 9, 126
foetus 130, 132
foramen ovale 132
fovea 114
fractionation 10
free radical 90
freeze fracture 7
fructose 16
FSH (follicle stimulating hormone) 127,
    131

fungus 52, 93

G

galactose 16, 23
gamete 39
gene 39
gene therapy 39
genotype 34, 44, 53
genus 51
global warming 64
glomerulus 107
glucagon 104, 119
glucose 15, 16, 23, 25, 27, 71, 104,
    106, 119
glucose isomerase 23
glucose oxidase 21, 23
glucose tolerance test 104
glycerol 19, 21, 67, 71
glycocalyx 8, 11
glycogen 5, 8, 9, 15, 104, 106, 119,
    123
glycolysis 25, 27
glycoprotein 16, 100
glycosidic bond 15, 16
Golgi body 8
Graafian follicle 125, 127
granum 29
greenhouse effect 64
grey matter 120
gross primary production 60
growth 133
growth curve 133, 134
growth spurt 134
guanine 32

H

haemoglobin 18, 34, 83, 84, 85, 86,
    129
haemoglobin, foetal 85
haemophilia 46
haemopoiesis 83
haploid 43
Hardy-Weinberg equilibrium 39, 53
HCG (human chorionic gonadotrophin)
    129
heart 77, 82, 88, 89, 90
heart attack 90
heart disease 90
heart sounds 89
heparin 16
hepatic portal vein 69, 82, 105, 106
Hepatitis B 99
hepatocyte 105, 106
Hering-Breuer reflex 81
heroin 113
Herpes 109
heterozygote 44
HIV (human immunodeficiency virus)
    100
hole-in-the-heart 90
homeostasis 81, 87, 103, 104, 108
Homo erectus 138
Homo habilis 138
Homo sapiens 138
homogenate 10
homologous pair 41, 43
homozygote 44

hormone 106, 118
HRT (hormone replacement therapy) 135
human evolution 137, 138
human growth hormone 37
humoral response 95
Huntington's disease 44
hybrid vigour 47
hybridoma 98
hydrocarbon 19
hydrogen bond 15, 18
hydrogencarbonate 86
hydrolase 21
hydrolysis 15, 18, 21, 28
hydrophilic head 11
hydrophobic tail 11
hyperpolarisation 116
hypertension 91
hypha 101
hypothalamus 102, 103, 130
hypothermia 102, 135

I

ileum 66
immunity 96
immunoglobulin 131
inbreeding 47
independent assortment 43, 45
induced fit hypothesis 20
industrial melanism 49
infarction 90
inflammation 94, 95
influenza 99
inhibitor, competitive and non-competitive 22
inhibitor, reversible and irreversible 22
inspiration 78, 81
inspiratory reserve volume 80
insulation, thermal 19
insulin 37, 104
intercostal muscle 77, 78, 81
interneurone 120
interphase 42
iodine solution 17
iodopsin 117
iris 114
iron 73, 106
ischaemic heart disease 90, 91
Islets of langerhans 118
isolation, reproductive 50

J

juxta-glomerular apparatus 108

K

karyotype 46
keratin 18
kidney 107
kinetochore 42
kingdom 51, 52
Klinefelter's syndrome 48
Kreb's cycle 26, 27
Kuppfer cell 106

L

labour 130
lactase 16, 23, 67

lactate dehydrogenase 21
lactate 25, 106
lactation 76, 131
lacteal 69
lactose 16, 23, 67, 76
laryngitis 77
larynx 77
LDL (low density lipoprotein) 91
legumes 73
lens 114
letdown reflex 131
leukocyte 8, 83
LH (luteinising hormone) 127, 131
ligament 136
ligase 21, 27
limiting factor 57, 59
Lincoln index 55
lipase 21, 67
lipid 14, 17, 19, 29, 71
liver 66, 68, 69, 78, 103, 104, 105, 106
lobule 105
lock-and-key hypothesis 20
longitudinal growth study 133
Loop of Henle 107
lung 39, 77, 78, 79
lymphatic system 69
lymphocyte 83, 93, 94, 95
lymphoid tissue 134
lysosome 8
lysozyme 9

M

Mab (monoclonal antibody) 98
macrophage 83, 91, 106
malaria 99
maltase 67
maltose 16, 67
mammal 52
mammary gland 131, 132
mark-recapture 55
mean 48
medulla 81, 92
medulla, of kidney 107
meiosis 43, 124, 125
melanin 34
membrane 5, 8, 19
membrane, selectively permeable 12
memory cell 95, 96
Mendel 45
menopause 135
menstrual cycle 127
menstruation 127
mesentery 68
mesoderm 128
mesosome 9
metabolic rate 74, 75
metabolism 5, 8, 26, 75, 76, 106, 135
metaphase 42, 43
methane 64
MHC (major histocompatibility complex) 95
microbody 8
microdiet 74
microfilaments 8
microtubule 8
microvilli 8, 69, 105

micturition 121
milk, breast 76
minisatellite 38
mitochondrion 8, 25, 27, 86, 111
mitosis 42, 128
mode 48
mollusc 52
monocyte 83
monohybrid 44
monosaccharide 15
morphine 81, 113
morula 128
motor neurone 109, 120
movement 136
mucopolysaccharide 16
mucosa 68
mutagen 34
mutation 34, 48, 101
myelin sheath 109, 112
myelin 19
myocarditis 90
myocardium 88
myoglobin 85
myosin 18, 28, 122

N

NAD (nicotinamide adenine dinucleotide) 20, 25, 26, 27
NADP (nicotinamide adenine dinucleotide phosphate) 30, 31
Neanderthal man 138
negative feedback 81, 102, 103, 104, 108
neo-Darwinism 49
nephron 107, 108
net primary production 60
neurilemma 109
neurotransmitter 111, 113, 116
niacin 20
niche 56
nicotine 111, 113
night vision 115
nitrate 63
nitrification 63
Nitrobacter 63
nitrogen cycle 63
nitrogen fixation 28, 63
Nitrosomonas 63
Node of Ranvier 109, 112
nomogram 75
non-disjunction 48
non-reducing sugar 17
nuclear pore 40
nucleic acid 14
nucleolus 40
nucleoplasm 40
nucleosome 40
nucleotide 14, 32, 41
nucleus 5, 8, 34, 36, 40, 126

O

obesity 90
objective lens 6, 7
oesophagus 66
oestrogen 19, 118, 127, 129, 130, 131
oil 17, 19
oilseed 73